W0106697

Advances in Numerical Mathematics

Angela Kunoth

Wavelet Methods –
Elliptic Boundary Value Problems
and Control Problems

Advances in Numerical Mathematics

Editors Hans Georg Bock Wolfgang Hackbusch
Mitchell Luskin Rolf Rannacher

Thomas Apel *Anisotropic Finite Elements: Local Estimates and Applications*

Jürgen Bey *Finite-Volumen- und Mehrgitter-Verfahren für elliptische Randwertprobleme*

Klaus Johannsen *Robuste Mehrgitterverfahren für die Konvektions-Diffusions-Gleichung mit wirbelbehafteter Konvektion*

Ralf Kornhuber *Adaptive Monotone Multigrid Methods for Nonlinear Variational Problems*

Andreas Prohl *Projection and Quasi-Compressibility Methods for Solving the Incompressible Navier-Stokes-Equations*

Reinhold Schneider *Multiskalen- und Wavelet-Matrixkompression Analysisbasierte Methoden zur effizienten Lösung großer vollbesetzter Gleichungssysteme*

Rüdiger Verfürth *A Review of A Posteriori Error Estimation and Adaptive Mesh-Refinement Techniques*

Wavelet Methods – Elliptic Boundary Value Problems and Control Problems

Von Prof. Dr. rer. nat. Angela Kunoth
Universität Bonn

B. G. Teubner Stuttgart · Leipzig · Wiesbaden

Prof. Dr. rer. nat. Angela Kunoth

1963 geboren in Emsdetten/Westfalen. Von 1982 bis 1990 Studium der Mathematik an der Universität Bielefeld, Diplom 1990. Promotion an der Freien Universität Berlin 1994. Wissenschaftliche Mitarbeiterin am Weierstrass-Institut für Angewandte Analysis und Stochastik in Berlin von 1994–1997 und am Institut für Angewandte und Praktische Mathematik der RWTH Aachen von 1997–1999. Habiltation in Mathematik, RWTH Aachen 2000. Längere Auslandsaufenthalte an der Univerity of South Carolina (1990/91), Sintef, Oslo (1994) und Texas A&M University (1995). Seit 1999 Professorin am Institut für Angewandte Mathematik der Universität Bonn.

Die Deutsche Bibliothek – CIP-Einheitsaufnahme
Ein Titeldatensatz für diese Publikation ist bei
Der Deutschen Bibliothek erhältlich.

1. Auflage April 2001

Der Verlag Teubner ist ein Unternehmen der Fachverlagsgruppe BertelsmannSpringer.

www.teubner.de

Gedruckt auf säurefreiem Papier
Umschlaggestaltung: Peter Pfitz, Stuttgart

ISBN-13:978-3-519-00327-4 e-ISBN-13:978-3-322-80027-5
DOI: 10.1007/978-3-322-80027-5

Abstract

This monograph is concerned with various aspects of wavelet concepts for the numerical solution of a class of minimization problems involving *elliptic partial differential equations.* Particular emphasis is placed on the treatment of the *boundary* and the *boundary conditions.* The boundary conditions are appended by Lagrange multipliers which leads to a *saddle point problem.* This approach is combined with a *fictitious domain method* into which the underlying domain is embedded.

In the first part of this monograph, such saddle point problems stemming from *elliptic boundary value problems* in Galerkin formulation are systematically investigated. The corresponding results constitute a foundation for a more involved problem, namely, to apply the approach to *control problems* with boundary control where the constraints are given in terms of an elliptic boundary value problem. Minimizing a functional involving the control and the solution of the boundary value problem leads to a *coupled system of saddle point problems.* For its numerical solution different iterative schemes are proposed and convergence is shown, which can be reduced to iterative schemes for saddle point problems. Moreover, one iterative method is proved to be asymptotically optimal.

The central tool for the numerical analysis of these classes of problems are *wavelets* which provide *norm equivalences* between function space norms and sequence norms: a function measured in a function space norm can be characterized in terms of a weighted sequence norm involving its coefficients in a *wavelet expansion.* This is used here to derive a *discretization concept* for a whole class of linear operators. This concept is described initially in an infinite–dimensional abstract setting, with particular emphasis on the flexible and adequate treatment of boundaries and boundary conditions. An essential ingredient is to guarantee that these operators, assembled from the differential and boundary operators, are isomorphisms on the relevant (energy) function spaces. By means of the wavelet setting, they are then in a second step transformed into ℓ_2–operators, still on an infinite–dimensional space. In a third step, conditions are derived which ensure that in the finite–dimensional case the discretizations are *stable*, implying that certain finite sections of these operators are uniformly bounded. Since these operators are then trivially again ℓ_2–operators, they have condition numbers which are *independent* of the discretization step size. This, in turn, entails that the speed of any iterative convergent method used for the solution of the linear systems does *not* deteriorate as the discretization step size becomes finer.

There are different ways of assuring stability of discretizations for saddle point problems. One involving a Ladyšenskaja–Babuška–Brezzi (LBB) condition is systematically investigated here for general domains in $\mathbb{R}^n$, requiring a certain choice of discretizations of the involved function spaces. A completely different path is followed by formulating the elliptic boundary value problem in saddle point form as a *least squares problem.* Corresponding results based on appropriately *truncating* wavelet expansions are presented here.

Since all these results rely to a great extent on the availability of appropriate wavelet bases on the domains and boundaries under consideration, the main facts concerning suitable constructions of wavelets are recalled from [CTU1, DKU2, DS3].

The different approaches are supplemented each by numerical examples.

Preface

While wavelets have since their discovery mainly been applied to problems in signal analysis and image compression, their analytic power has more and more also been recognized for problems in Numerical Analysis. Together with the functional analytic framework for different differential and integral quations, one has been able to conceptually discuss questions which are relevant for the fast numerical solution of such problems: preconditioning issues, derivation of stable discretizations, compression of fully populated matrices, evaluation of non-integer or negative norms, and adaptive refinements based on A-posteriori error estimators.

This research monograph focusses on applying wavelet methods to elliptic differential equations. Particular emphasis is placed on the treatment of the boundary and the boundary conditions. Moreover, a control problem with an elliptic boundary problem as contraint serves as an example to show the conceptual strengths of wavelet techniques for some of the above mentioned issues.

At this point, I would like to express my gratitude to several people before and during the process of writing this monograph. Most of all, I wish to thank Prof. Dr. Wolfgang Dahmen to whom I personally owe very much and with whom I have co–authored a large part of my work. He is responsible for the very stimulating and challenging scientific atmosphere at the Institut für Geometrie und Praktische Mathematik, RWTH Aachen. We also had an enjoyable collaboration with Prof. Dr. Reinhold Schneider from the Technical University of Chemnitz. Here I take the opportunity to thank him very much for the inspiring discussions. I express my sincere gratitude to Prof. Dr. Wolfgang Dahmen, Prof. Dr. Claudio Canuto from the Politecnico di Torino and Prof. Dr. Henning Esser from Aachen for various remarks and improvements of the material in this monograph. I am especially indebted for their detailed comments, valuable suggestions and various improvements. For discussions on theoretical issues I wish to thank Dr. Luise Blank, PD Dr. Stephan Dahlke and Dr. Karsten Urban. The software package `IGPMLib, Version 2.0`, which has been developed at the IGPM has been used as a basis to generate the numerical results in this thesis. For their support and help with programming issues, I am very grateful to Arne Barinka, Dr. Titus Barsch, Frank Knoben, Dr. Karsten Urban and Jürgen Vorloeper.

Bonn, February 2001 Angela Kunoth

Contents

1 Introduction

During the past decade the use of different types of wavelets has to a large extent been dominated by applications to signal analysis and image compression problems. Their potential in numerical schemes for various operator equations has also aroused increasing interest [C, D3]. This is largely due to the fact that *wavelet bases* or shortly *wavelets* combine at least three useful properties, namely,

(I) they constitute a *Riesz basis* for the relevant function space such as an energy space;

(II) they can be arranged to be *compactly supported*;

(III) they satisfy moment conditions up to a certain order.

Thus, wavelets are principally able to decompose an object into local components which are characterized by different length scales. Consequently, this allows for approximations of such an object with respect to different *resolutions* including adapting the resolution locally. The key feature of wavelets is the *combination* of properties (I)–(III), making them stand out over other basis–oriented discretization schemes for operator equations. For differential operators, essentially properties (I) and (II) are used. The importance of the third property is here, in contrast to integral operators [DPS1], not so apparent, though implicitly contained in (I). Properties (I)–(III) are, in particular, satisfied by the most well–known example, the Haar wavelets, which constitute an L_2–orthonormal basis of compact support for $L_2(\mathbb{R})$. The enormous interest for wavelets themselves came up with the construction of smoother orthogonal variants with compact support, the Daubechies wavelets [Dau]. As it turned out, *biorthogonal wavelet bases* [CDF] are sufficient to guarantee properties (I)–(III) and offer the most flexibility to allow for compact support and symmetry of the primal and dual functions which are essential features for computations [D2]. For the treatment of operator equations, a pivotal role is taken by the biorthogonal wavelets which are generated by *B–Splines*.

The above listed properties of wavelets will be used as follows. Firstly, a *general concept* for the treatment of linear operator equations will be described in **Chapter 2**. The starting point is a general *system* of (weakly defined) operator equations

$$\mathcal{L}U = F \tag{1.1}$$

which is assumed to establish an *isomorphism* from a certain product space $\mathcal{H}$ into its dual $\mathcal{H}'$,

$$\|\mathcal{L}V\|_{\mathcal{H}'} \sim \|V\|_{\mathcal{H}}, \qquad V \in \mathcal{H}. \tag{1.2}$$

Here the notation $a \sim b$ always means that a can be estimated from below and from above by a constant multiple of b. For example, in the Lagrange multiplier approach for appending essential boundary conditions discussed in Chapter 4, $\mathcal{L}$ is a saddle point operator and $\mathcal{H}$ a product of Sobolev spaces on the domain of the problem and on its boundary. In the context of control problems discussed in Chapter 6, $\mathcal{L}$ contains blocks of saddle point operators.

Assuming the existence of a wavelet basis for $\mathcal{H}$, property (I) is then used to transform the system of continuous equations (1.1) into discrete infinite-dimensional ones,

$$\mathbf{L}\,\mathbf{U} = \mathbf{F} \tag{1.3}$$

where $\mathbf{L}$ is an isomorphism from $\ell_2(\mathbb{I})$ to $\ell_2(\mathbb{I})$. $\mathbb{I}$ always stands for an infinite index set whose indices comprise different information such as scale and spatial location. In other words, problem (1.3) is *well-posed in Euclidean metric.*

The second cornerstone of the common platform is to find conditions that guarantee *stable discretization schemes* for (1.3). That is, certain finite sections of $\mathbf{L}$ with respect to a *finite* index set $\Lambda \subset \mathbb{I}$ also provide an isomorphism on $\ell_2(\Lambda)$ with constants in the norm equivalence which do *not* depend on Λ,

$$\|\mathbf{L}_\Lambda\, \mathbf{V}_\Lambda\|_{\ell_2(\Lambda)} \sim \|\mathbf{V}_\Lambda\|_{\ell_2(\Lambda)}. \tag{1.4}$$

This means that the stiffness matrix in wavelet basis $\mathbf{L}_\Lambda$ has *uniformly bounded* condition numbers. This entails e.g. for symmetric positive definite $\mathbf{L}_\Lambda$ that the *convergence speed* of an appropriate iterative scheme like the conjugate gradient method for the solution of $\mathbf{L}_\Lambda \mathbf{U}_\Lambda = \mathbf{F}_\Lambda$ is *independent* of Λ.

The results in this monograph rely to a great extent on the availability of appropriate wavelet bases on the domains and manifolds under consideration. Thus, some effort has been made to provide sufficient general constructions of such wavelets satisfying (I)—(III). Starting with biorthogonal wavelet bases on $\mathbb{R}$ from [CDF], one can construct first biorthogonal wavelets on the unit interval [DKU2], and then use tensor products to obtain biorthogonal wavelets on cubes in $\mathbb{R}^n$. We recall in **Chapter 3** the necessary results from [DKU2]. Furthermore, for domains or boundaries which are unions of parametric images of cubes, we collect the major ideas from [DS1, DS3].

The remaining three chapters of this monograph are devoted to establishing (1.2) for different systems of operator equations (1.1) involving elliptic boundary value problems, to deriving stable wavelet-based discretizations, to discussing solution techniques of the resulting systems of linear equations, and to support the theoretical results by providing corresponding numerical examples. In pursuing this plan, different obstructions are encountered that have to be tackled.

Chapter 4 deals with elliptic boundary value problems and the explicit treatment of essential boundary conditions (on the whole boundary or on part of it). *Appending boundary conditions* by Lagrange multipliers [Bal] leads to a saddle point problem. This method is combined with a *fictitious domain* approach where the original domain Ω is embedded into a larger simple one denoted by $\square$. The advantage of this method is that the differential operator on the one hand and the boundary and the boundary operator on the other hand are to some extent separated.

The resulting combined *Fictitious Domain—Lagrange Multiplier Approach* leads to a linear operator equation of the following form: Given $(f,u) \in (H^1(\square))' \times H^{1/2}(\Gamma)$, find $(y,p) \in H^1(\square) \times (H^{1/2}(\Gamma))'$ such that

$$\mathcal{L}\begin{pmatrix} y \\ p \end{pmatrix} := \begin{pmatrix} A & B' \\ B & 0 \end{pmatrix}\begin{pmatrix} y \\ p \end{pmatrix} = \begin{pmatrix} f \\ u \end{pmatrix}. \tag{1.5}$$

Here Γ is (part of) the boundary of Ω, B represents the *trace operator* and A the elliptic differential operator. The trace theorem which connects the space on the domain, $H^1(\Box)$, with the one on the boundary, $(H^{1/2}(\Gamma))'$, implies the *inf–sup condition* for the saddle point problem (1.5) and, thus, ensures that the resulting saddle point operator on the continuous spaces is an isomorphism. The Fictitious Domain—Lagrange Multiplier Approach facilitates that for the discretized saddle point systems the main and most time–consuming part of the work can be carried out on a simple domain for which one can use fast efficient solvers. The boundary and the boundary conditions are taken care of by an operator on a lower dimensional manifold, thus, involving in the discrete case much less degrees of freedom.

This Fictitious Domain—Lagrange Multiplier Approach is particularly suited for problems where boundary conditions are frequently changing. An instance for the latter situation are the *control problems* treated in Chapter 6 where the control is exerted through the boundary conditions. Here the techniques discussed in Chapter 4 are extended to fit these situations.

For the saddle point problem (1.5), it remains to ensure that the discrete finite–dimensional scheme is stable. The counterpart to the inf–sup condition for the discrete spaces is the Ladyšenskaja–Babuška–Brezzi (LBB) condition. For general domains in $\mathbb{R}^n$, criteria for its validity are derived in Section 4.2.4 using techniques from approximation theory. An essential feature of the results is that the discretization spaces on the boundary Γ are *not* (subsets of) trace spaces of the ones on $\Box$.

An interesting alternative to Galerkin discretizations for saddle point problems which avoids the constraints imposed by the LBB condition is offered by certain *least squares formulations* based on the dual norms $\|\cdot\|_{\mathcal{H}'}$ in (1.2). These dual norms are often norms for Sobolev spaces with negative or noninteger indices such as the H^{-1}– or the $H^{1/2}$–norm. While in the context of finite element methods, efficient preconditioning techniques have revived least squares approaches during the past years, see e.g. [BLP2, CLMM], the evaluation of such dual norms has been a major obstacle. Here the main guide line has been to avoid too stringent regularity requirements on the solution imposed by an improper choice of the least squares functional, see e.g. [BLP1, BLP2, PCL]. The practical feasibility of these formulations in the wavelet framework hinges, among other things, again on (I). In fact, this approach is *not* confined to saddle point problems but applies as long as (1.2) holds and corresponding norm equivalences are available. This is described in **Chapter 5**, see also [DKS2]. Some numerical results conclude this chapter.

Finally, in **Chapter 6** a class of problems is investigated where the treatment of boundary conditions by Lagrange multipliers is the method of choice because they vary in an extremal problem. It deals with *control problems* of the following type. A quadratic functional involving the *natural* norms of the solution of an elliptic boundary value problem and its boundary control is to be minimized.

A simple example of such a control problem is the following. Let the domain be $\Box = \Omega = (0,1)^2$ with boundary $\partial\Omega$. The right face of $\partial\Omega$ is labelled as Γ_y where values for y are prescribed. The opposite face is denoted as Γ, the 'control boundary', see Figure 1.1. Now the control problem can be formulated as follows: Given $f \in (H^1(\Omega))'$, find the solution $(y,p) \in H^1(\Omega) \times (H^{1/2}(\Gamma))'$ of the elliptic boundary value problem in weak form (1.5),

$$\begin{pmatrix} A & B' \\ B & 0 \end{pmatrix} \begin{pmatrix} y \\ p \end{pmatrix} = \begin{pmatrix} f \\ u \end{pmatrix}, \tag{1.6}$$

such that for a given $y_{\Gamma_y} \in H^{1/2}(\Gamma_y)$ the functional

$$\mathcal{J}(y,u) \;=\; \tfrac{1}{2}\,\|y - y_{\Gamma_y}\|^2_{H^{1/2}(\Gamma_y)} + \tfrac{1}{2}\,\|u\|^2_{H^{1/2}(\Gamma)} \tag{1.7}$$

is minimized. The function u acting on Γ is the *boundary control* for the system (1.6).

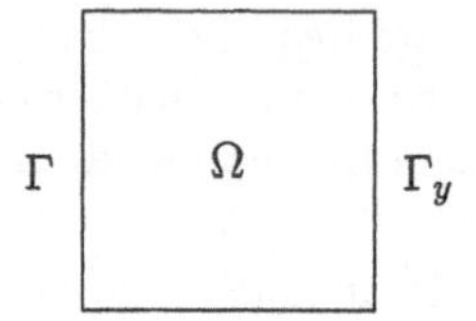

Figure 1.1: Domain and boundaries for the control problem.

In the numerical treatment of such control problems one faces several difficulties. One difficulty is the evaluation of fractional norms in (1.7) which naturally appear. In finite element approaches, these norms are avoided by either taking the H^1 or the L_2 norm which then possibly requires to assume higher regularity. The minimization of (1.7) can be shown to be equivalent to a weakly coupled system of two saddle point problems. Here for large–scale problems complexity issues and then appropriate solution methods have to be considered. This type of problems provides a promising potential for the use of wavelets: from the evaluation of non–integer norms over efficient preconditioning techniques to the particular treatment of essential non–homogeneous boundary conditions.

In Chapter 6 such control problems are fitted into the general concept of Chapter 2. The resulting coupled system is proved to be an isomorphism so that all the previous strategies can be applied. For the numerical solution of the coupled saddle point problems, one can use the strategy consisting of an outer iteration where in each iteration step the two saddle point problems are solved alternately. Convergence of outer gradient schemes seems to have been proved so far only when direct methods like QR decomposition have been employed for the solution of each of the saddle point problems. This severely constrains the size of such problems and essentially excludes realistic applications in several space variables. Here the employment of efficient iterative solvers for such large coupled systems of equations based on the wavelet concepts from the previous chapters is investigated. Aside from a least squares formulation using the material from Chapter 5, different iterative strategies are discussed. In particular, a *fully iterative method* is proposed here which is shown to be *asymptotically optimal.*

Throughout this monograph, important constants are written explicitly. If the precise value of a constant does not matter, the notation $a \lesssim b$ or $b \gtrsim a$ is used to express that a can be bounded by a constant multiple of b, and $a \sim b$ means $a \lesssim b$ and $b \lesssim a$.

2 The General Concept

The minimization problems that are treated in this monograph can be rephrased as systems of operator equations that fit into the following abstract framework.

We consider certain closed subspaces $H_{i,0}$, $i = 1, \ldots, M$, of Hilbert spaces H_i whose inner products are denoted by $(\cdot,\cdot)_{H_i}$. The space $H'_{i,0}$ is the normed dual of $H_{i,0}$ endowed with the norm

$$\|v\|_{H'_{i,0}} := \sup_{0 \neq w \in H_{i,0}} \frac{\langle w, v\rangle}{\|w\|_{H_i}} \tag{2.1}$$

where $\langle\cdot,\cdot\rangle$ always denotes the (respective) dual form. The H_i will be Sobolev spaces on either the domain Ω or $\Box$, or on boundary manifolds $\partial\Omega$ or smooth subsets Γ of $\partial\Omega$. Thus, we have

$$\text{either} \qquad H_i \subseteq L_2 \subseteq H'_{i,0} \qquad \text{or} \qquad H'_{i,0} \subseteq L_2 \subseteq H_i, \tag{2.2}$$

where L_2 stands for the space of Lebesgue integrable functions $L_2 = L_2(\Omega)$, $L_2(\Box)$, or $L_2 = L_2(\partial\Omega)$, $L_2(\Gamma)$. Furthermore, the spaces $H_{i,0}$ will either coincide with H_i or will be determined by homogeneous boundary conditions.

Given $H_{i,0}$, $i = 1, \ldots, M$, we define the product space

$$\mathcal{H} := \prod_{i=1}^{M} H_{i,0}, \qquad \text{where} \quad (\cdot,\cdot)_{\mathcal{H}} := \sum_{i=1}^{M} (\cdot,\cdot)_{H_i} \tag{2.3}$$

is the canonical inner product on $\mathcal{H}$. Thus, the norms for $\mathcal{H}$ and $\mathcal{H}'$ can be defined as

$$\|V\|_{\mathcal{H}} := \left(\sum_{i=1}^{M} \|v_i\|^2_{H_i}\right)^{1/2}, \qquad \|W\|_{\mathcal{H}'} = \sup_{\|V\|_{\mathcal{H}}=1} \langle V, W\rangle, \tag{2.4}$$

where $\langle V, W\rangle := \sum_{i=1}^{M} \langle v_i, w_i\rangle$.

Step 1 — Well–Posedness:
The starting point is an important aspect in modeling the problem. The task is to formulate the problem in question in such a way that one can *identify* $\mathcal{H}$ and a linear operator $\mathcal{L}$ such that $\mathcal{L} : \mathcal{H} \to \mathcal{H}'$ is an *isomorphism*.

To this end, suppose that $A_{i,\ell}(\cdot,\cdot)$ are bounded bilinear forms on $H_i \times H_\ell$,

$$A_{i,\ell}(v, w) \lesssim \|v\|_{H_i}\|w\|_{H_\ell}, \quad i, \ell = 1, \ldots, M, \tag{2.5}$$

and define for $V = (v_1, \ldots, v_M)^T \in \mathcal{H}$

$$A_i(w, V) := \sum_{\ell=1}^{M} A_{i,\ell}(w, v_\ell), \quad w \in H_{i,0}, \quad i = 1, \ldots, M. \tag{2.6}$$

The variational problem we will be concerned with is the following: Given $f_i \in H'_{i,0}$ for $i = 1, \ldots, M$, find $U \in \mathcal{H}$ such that

$$A_i(w, U) = \langle w, f_i\rangle, \quad w \in H_{i,0}, \quad i = 1, \ldots, M. \tag{2.7}$$

The crucial property that will have to be verified in each concrete example is that (2.7) is well–posed in the following sense. Let the operators

$$\mathcal{L}_{i,\ell} : H_{\ell,0} \to H'_{i,0}, \quad \mathcal{L}_i : \mathcal{H} \to H'_{i,0} \tag{2.8}$$

be defined by

$$\langle w, \mathcal{L}_{i,\ell} v\rangle = A_{i,\ell}(w,v), \quad \langle w, \mathcal{L}_i U\rangle := A_i(w,U), \quad w \in H_{i,0}. \tag{2.9}$$

Thus, (2.7) is equivalent to the following operator equation: Given $F = (f_1, \ldots, f_M)^T \in \mathcal{H}'$, find $U = (u_1, \ldots, u_M)^T \in \mathcal{H}$ such that

$$\sum_{j=1}^{M} \mathcal{L}_{i,j} u_j =: \mathcal{L}_i U = f_i, \quad i = 1, \ldots, M. \tag{2.10}$$

Briefly this reads

$$\mathcal{L}U = F, \tag{2.11}$$

where $\mathcal{L} = (\mathcal{L}_{i,j})_{i,j=1}^M$.

Well–posed now means that $\mathcal{L}$ is an isomorphism from $\mathcal{H}$ to $\mathcal{H}'$, i.e.,

$$\sum_{i=1}^{M} \|v_i\|_{H_i}^2 \sim \sum_{i=1}^{M} \|\mathcal{L}_i V\|_{H'_{i,0}}^2, \quad V \in \mathcal{H}, \tag{2.12}$$

or briefly

$$\|\mathcal{L} V\|_{\mathcal{H}'} \sim \|V\|_{\mathcal{H}} \qquad \text{for all } V \in \mathcal{H}. \tag{2.13}$$

Specifically this means that there exist constants $0 < c_{\mathcal{L}} \leq C_{\mathcal{L}} < \infty$ such that

$$c_{\mathcal{L}}\|V\|_{\mathcal{H}} \leq \|\mathcal{L}V\|_{\mathcal{H}'} \leq C_{\mathcal{L}}\|V\|_{\mathcal{H}}, \quad V \in \mathcal{H}, \tag{2.14}$$

holds.

Various instances including different formulations of second order elliptic boundary value problems or transmission problems are summarized in [DKS2] where $\mathcal{H}$ and $\mathcal{L}$ with the above properties are identified.

Step 2 — Shifting:
Once Step 1 is established, wavelet concepts can be employed to transform (2.11) into an equivalent well-posed ℓ_2-*problem*, see [D3].

To this end, define $\mathbb{I} := \mathbb{I}_1 \times \cdots \times \mathbb{I}_M$ with infinite index sets $\mathbb{I}_i$. For each i, the elements λ of $\mathbb{I}_i$ consist of different types of indices such as the *level of resolution* (*refinement* or *discretization level*) denoted by $|\lambda|$ and the *spatial location*. What we call *wavelets* for $\mathcal{H} = H_{1,0} \times \cdots \times H_{M,0}$ is a (catenated) collection of functions

$$\Psi := \{{}^1\Psi, \ldots, {}^M\Psi\} \tag{2.15}$$

where for each $i = 1, \ldots, M$ the collection ${}^i\Psi$ defined by

$${}^i\Psi := \{{}^i\psi_\lambda : \lambda \in \mathbb{I}_i\} \subset H_{i,0} \tag{2.16}$$

has the following properties:

(I) *Riesz basis property:* Every function v in $H_{i,0}$ can be uniquely expanded in terms of ${}^{i}\Psi$,

$$v = \mathbf{v}^T\,{}^{i}\Psi := \sum_{\lambda \in I\!I_i} v_\lambda\,{}^{i}\psi_\lambda, \tag{2.17}$$

and its expansion coefficients satisfy the *norm equivalence*

$$\|v\|_{H_i} \sim \|{}^{i}\mathbf{D}\mathbf{v}\|_{\ell_2(I\!I_i)} \tag{2.18}$$

where ${}^{i}\mathbf{D}$ is some diagonal matrix. In other words, the *scaled* collection ${}^{i}\mathbf{D}^{-1}({}^{i}\Psi)$ constitutes a *Riesz basis* for $H_{i,0}$. We always use the convention that the basis is normalized in L_2, that is, when H_i agrees with L_2 it follows that ${}^{i}\mathbf{D} = \mathbf{I}$.

(II) *Locality:* The wavelets ${}^{i}\psi_\lambda$ are compactly supported with decreasing support width when the discretization level $|\lambda|$ grows,

$$\mathrm{diam}\,(\mathrm{supp}\,{}^{i}\psi_\lambda) \sim 2^{-|\lambda|}. \tag{2.19}$$

With the aid of Riesz' representation theorem, one can conclude that for the dual pairing $\langle\cdot,\cdot\rangle$ for $H_{i,0}$ and its dual $H'_{i,0}$ there exists a collection

$${}^{i}\tilde{\Psi} := \{{}^{i}\tilde{\psi}_\lambda : \lambda \in I\!I_i\} \subset H'_{i,0} \tag{2.20}$$

such that

$$\langle {}^{i}\psi_\lambda, {}^{i}\tilde{\psi}_\mu\rangle = \delta_{\lambda\mu}, \qquad \lambda, \mu \in I\!I_i, \tag{2.21}$$

and ${}^{i}\mathbf{D}({}^{i}\tilde{\Psi})$ is a Riesz basis for $H'_{i,0}$. Here $\delta_{\lambda\mu}$ is the Kronecker delta. In fact, by a duality argument one concludes from (2.18) that the corresponding norm equivalence

$$\|\tilde{v}\|_{H'_{i,0}} \sim \|{}^{i}\mathbf{D}^{-1}\tilde{\mathbf{v}}\|_{\ell_2(I\!I_i)} \tag{2.22}$$

holds for any $\tilde{v} = \tilde{\mathbf{v}}^T\,{}^{i}\tilde{\Psi} \in H'_{i,0}$ [D2]. The coefficients v_λ in the expansion (2.17) can then be expressed in terms of the dual basis as $v_\lambda = \langle v, {}^{i}\tilde{\psi}_\lambda\rangle$.

$({}^{i}\Psi, {}^{i}\tilde{\Psi})$ is called a pair of *biorthogonal wavelets.* Of particular interest are the cases when the dual wavelets ${}^{i}\tilde{\Psi}$ also have compact support (2.19).

Shorthand Notation:
In the remainder of this monograph, the following shorthand notation will be used. We will view ${}^{i}\Psi$ and Ψ as in (2.16), (2.15) as a *collection* of functions as well as a (possibly infinite) (column) *vector* containing all functions always assembled in some fixed unspecified order. For a countable collection of functions Θ and some single function σ, the quantities $\langle\Theta,\sigma\rangle$ and $\langle\sigma,\Theta\rangle$ are to be understood as the column, respectively row, vector with entries $\langle\theta,\sigma\rangle$, respectively $\langle\sigma,\theta\rangle$, $\theta \in \Theta$. For two collections Θ, Σ, the term $\langle\Theta,\Sigma\rangle$ is then a (possibly infinite) matrix with entries $(\langle\theta,\sigma\rangle)_{\theta\in\Theta,\,\sigma\in\Sigma}$ for which $\langle\Theta,\Sigma\rangle = \langle\Sigma,\Theta\rangle^T$. This also implies for a (possibly infinite) matrix $\mathbf{C}$ that $\langle\mathbf{C}\Theta,\Sigma\rangle = \mathbf{C}\langle\Theta,\Sigma\rangle$ and $\langle\Theta,\mathbf{C}\Sigma\rangle = \langle\Theta,\Sigma\rangle\mathbf{C}^T$.

In this notation, the expansion coefficients in (2.17) and (2.22) can explicitly be expressed as

$$\mathbf{v}^T = \langle v, {}^{i}\tilde{\Psi} \rangle, \qquad \tilde{\mathbf{v}} = \langle {}^{i}\Psi, \tilde{v} \rangle. \tag{2.23}$$

Note that the *biorthogonality* or *duality conditions* (2.21) can now be written in terms of an infinite matrix,

$$\langle {}^{i}\Psi, {}^{i}\tilde{\Psi} \rangle = \mathbf{I}, \tag{2.24}$$

where $\mathbf{I}$ is the identity matrix.

Now we are ready to transform the continuous operator equation (2.11) into a discrete system of equations in terms of the wavelet basis Ψ for $\mathcal{H}$ as follows. Expansion of the solution

$$U = (u_1, \ldots, u_M)^T = \left(\mathbf{u}_1^T \, ({}^1\mathbf{D}^{-1}) \, {}^1\Psi, \ldots, \mathbf{u}_M^T \, ({}^M\mathbf{D}^{-1}) \, {}^M\Psi\right)^T =: \mathbf{U}^T \mathbf{D}^{-1} \Psi \quad \in \mathcal{H} \tag{2.25}$$

and the right hand side $F = \langle \Psi, F \rangle^T \tilde{\Psi} \in \mathcal{H}'$ yields the system of equations

$$\langle \Psi, \mathcal{L}\Psi \rangle \, \mathbf{D}^{-1} \, \mathbf{U} = \langle \Psi, F \rangle. \tag{2.26}$$

Multiplying the system (2.26) by the diagonal matrix $\mathbf{D}^{-1}$,

$$\mathbf{D}^{-1} \langle \Psi, \mathcal{L}\Psi \rangle \, \mathbf{D}^{-1} \, \mathbf{U} = \mathbf{D}^{-1} \langle \Psi, F \rangle, \tag{2.27}$$

and recalling the norm equivalences (2.18) and (2.22), we obtain

$$\mathbf{L} \, \mathbf{U} = \mathbf{F} \tag{2.28}$$

where $\mathbf{L} : \ell_2(I\!I) \to \ell_2(I\!I)$ is an isomorphism. Here we have used the abbreviations

$$\mathbf{L} := \mathbf{D}^{-1} \langle \Psi, \mathcal{L}\Psi \rangle \, \mathbf{D}^{-1}, \quad \mathbf{F} := \mathbf{D}^{-1} \langle \Psi, F \rangle. \tag{2.29}$$

In summary, the wavelet framework provides on account of the norm equivalences (2.18) and (2.22) a mechanism that allows to transform the original operator equation (2.11) into an equivalent infinite linear system of equations (2.28) which is well-posed in the Euclidean metric $\ell_2(I\!I)$. This implies also that the *spectral condition number* of $\mathbf{L}$ satisfies

$$\kappa(\mathbf{L}) := \|\mathbf{L}\|_2 \, \|\mathbf{L}^{-1}\|_2 \;\lesssim\; 1. \tag{2.30}$$

The constants on the right hand side are quotients of the constants in the isomorphism relation (2.13) and the norm equivalences (2.18) and (2.22), see e.g. [D3]. Thus, the multiplication by $\mathbf{D}^{-1}$ in (2.27) can be viewed as *preconditioning* the operator in (2.26).

Step 3 — Stability of the Discrete Systems:
The last cornerstone of the proposed strategy is to establish conditions that guarantee that in the finite-dimensional case the discretizations are *stable*. Depending on the situation, there are different ways to ensure this.

Galerkin Stability:
Let $\Lambda \subset I\!I$ be the product of any *finite* index sets Λ_i, $i = 1, \ldots, M$, and define

$$\Psi_\Lambda := \{\Psi_\lambda : \lambda \in \Lambda\} \tag{2.31}$$

and correspondingly $\tilde{\Psi}_\Lambda$. A *Galerkin scheme* for (2.11) is to find $U_\Lambda \in S(\Psi_\Lambda) := \mathrm{span}\{\Psi_\Lambda\}$ such that

$$\langle \psi_\lambda, \mathcal{L} U_\Lambda \rangle \;=\; \langle \psi_\lambda, F \rangle, \qquad \psi_\lambda \in \Psi_\Lambda, \tag{2.32}$$

holds. In order to describe what stability means in this context, define the projectors $P_\Lambda : \mathcal{H} \to S(\Psi_\Lambda)$, $P'_\Lambda : \mathcal{H}' \to S(\tilde{\Psi}_\Lambda)$,

$$P_\Lambda V := \langle V, \tilde{\Psi}_\Lambda \rangle \Psi_\Lambda, \qquad P'_\Lambda \tilde{V} := \langle \Psi_\Lambda, \tilde{V} \rangle^T \tilde{\Psi}_\Lambda. \tag{2.33}$$

Then the Galerkin scheme (2.32) is called *stable* if

$$\|P_\Lambda V\|_{\mathcal{H}} \;\sim\; \|P'_\Lambda \mathcal{L} P_\Lambda V\|_{\mathcal{H}'}, \qquad V \in \mathcal{H}, \tag{2.34}$$

holds *uniformly* in Λ. In terms of the bases $\Psi_\Lambda, \tilde{\Psi}_\Lambda$, the stability requirement (2.34) reads

$$\|V_\Lambda\|_{\mathcal{H}} \;\sim\; \|\langle \Psi_\Lambda, \mathcal{L} V_\Lambda \rangle^T \tilde{\Psi}_\Lambda\|_{\mathcal{H}'}, \qquad V_\Lambda \in S(\Psi_\Lambda). \tag{2.35}$$

Galerkin stability is also connected to the condition number of the involved linear operator. Expanding $U_\Lambda = \mathbf{U}_\Lambda^T \mathbf{D}_\Lambda^{-1} \Psi_\Lambda$ and applying the preconditioning (2.29), the system (2.32) is rewritten as

$$\mathbf{L}_\Lambda \mathbf{U}_\Lambda := \mathbf{D}_\Lambda^{-1} \langle \Psi_\Lambda, \mathcal{L} \Psi_\Lambda \rangle \mathbf{D}_\Lambda^{-1} \mathbf{U}_\Lambda \;=\; \mathbf{D}_\Lambda^{-1} \langle \Psi_\Lambda, F \rangle =: \mathbf{F}_\Lambda, \tag{2.36}$$

where $\mathbf{L}_\Lambda : \ell_2(\Lambda) \to \ell_2(\Lambda)$. Galerkin stability entails that $\mathbf{L}_\Lambda$ is still an isomorphism on $\ell_2(\Lambda)$ *uniformly* in Λ, implying that the spectral condition number of $\mathbf{L}_\Lambda$ is *bounded uniformly* in Λ,

$$\kappa(\mathbf{L}_\Lambda) = \|\mathbf{L}_\Lambda\|_2 \, \|\mathbf{L}_\Lambda^{-1}\|_2 \;\lesssim\; 1 \tag{2.37}$$

with constants consisting of the quotients of the constants from (2.35) and again the norm equivalences (2.18) and (2.22) but not depending on Λ.

An example where Galerkin stability holds are symmetric elliptic systems.

Additional Conditions Ensuring Stability:
One situation where additional conditions on the discretizations ensure stability are the saddle point problems treated in Chapter 4. There the LBB condition comes into play, see Section 4.2.4.

Least Squares Formulations:
Appropriate least squares formulations of (2.10) are an alternative to the previous situation which tends to become complicated when $M > 2$, i.e., when more than two approximation spaces need to be coupled. We will discuss conditions ensuring stability for a situation applying to *any* system of the above form (2.11) in Chapter 5. This strategy can in fact be interpreted as a Galerkin scheme for a *different* system of operator equations, leading to a *symmetric positive definite* system of linear equations.

Once Step 3 is completed, it remains to solve

$$\mathbf{L}_\Lambda \mathbf{U}_\Lambda = \mathbf{F}_\Lambda \tag{2.38}$$

numerically. For instance, for symmetric positive definite operators one can apply a conjugate gradient method which on account of (2.37) converges with a *constant* convergence speed independent of Λ. For the saddle point systems derived in Chapter 4, one can apply an Uzawa type algorithm which inherits this property with a proper scaling.

The general concept is briefly illustrated in Figure 2.1. It has been used already for different situations in [CDD1, DK1, DK2, DKS2, DPS1].

For the various types of operator equations treated in the next chapters, the following strategy will be pursued. For each of the different approaches, the general concept presented in this chapter will be applied. First in Step 1 the spaces $\mathcal{H}, \mathcal{H}'$ will be identified for which an appropriate $\mathcal{L}$ in (2.11) is an isomorphism (2.13) to establish that the problem is well-posed. For Step 2, different function spaces on domains or boundary manifolds will be used for which wavelet bases satisfying the necessary requirements collected in this chapter are needed. A corresponding construction of wavelets which can be used for these function spaces is recalled in Chapter 3. The remainder of the work consists then in establishing situations such that the finite-dimensional systems are stable in the sense of Step 3.

A few words on the indexing of the wavelets are in order. In order to describe the main features in Step 3, it has been convenient to use an index λ. In the next chapter, we will specify this index in the framework of uniform refinements as a tupel of indices (j, k) comprising information such as the level of resolution j and the location k. It may also contain information on the type of wavelet in more than one dimension. We will for convenience sometimes switch between both representations.

$$\mathcal{L}U = F$$

$$\mathcal{L} : \mathcal{H} \to \mathcal{H}' \quad \text{Isomorphism}$$

$$\|\mathcal{L}V\|_{\mathcal{H}'} \sim \|V\|_{\mathcal{H}}$$

Continuous Operator Equation

$\Downarrow$ Wavelet Bases for $\mathcal{H}$

$$\mathbf{L}\,\mathbf{U} = \mathbf{F}$$

$$\mathbf{L} : \ell_2(I\!I) \to \ell_2(I\!I) \quad \text{Isomorphism}$$

$$\|\mathbf{L}\mathbf{V}\|_{\ell_2(I\!I)} \sim \|\mathbf{V}\|_{\ell_2(I\!I)}$$

Discretized Infinite–Dimensional Operator Equation

$\Downarrow$ Stability of Discretizations

$\Lambda \subset I\!I, \quad \#\Lambda < \infty$

$$\mathbf{L}_\Lambda\,\mathbf{U}_\Lambda = \mathbf{F}_\Lambda$$

$$\mathbf{L}_\Lambda : \ell_2(\Lambda) \to \ell_2(\Lambda) \quad \text{Isomorphism}$$

$$\|\mathbf{L}_\Lambda \mathbf{V}_\Lambda\|_{\ell_2(\Lambda)} \sim \|\mathbf{V}_\Lambda\|_{\ell_2(\Lambda)}$$

Discretized Finite–Dimensional Operator Equation

Figure 2.1: General concept.

3 Wavelets

The general concept for the solution of operator equations described in Chapter 2 relies to a great extent on the availability of appropriate wavelet bases. During the past years, much effort has been invested in constructing various types of wavelets, including biorthogonal ones which are well suited for the numerical analysis of operator equations of the form (2.11). When speaking about 'wavelets', here always a pair of *biorthogonal wavelets* Ψ, $\tilde{\Psi}$ is meant, satisfying the Riesz bases, the locality and the duality properties (2.17), (2.18), (2.19), (2.22) and (2.24).

This chapter is devoted to a summary of the basic construction principles of such wavelets. Their cornerstones are multiresolution analyses of the relevant function spaces and the concept of stable completions which are recalled in Section 3.2. These concepts are free of Fourier techniques to allow for domains or manifolds which are subsets of $\mathbb{R}^n$ as they typically appear in the treatment of boundary value problems. Starting with biorthogonal wavelets on $\mathbb{R}$, a summary of the construction of such bases on an interval is given in Section 3.3. These bases are used in Section 3.4 to build biorthogonal wavelet bases on manifolds by means of tensor products, domain decomposition techniques and parametric mappings. But first a few remarks on the function spaces for which the wavelet bases will be needed are in order.

3.1 Preliminaries

The function spaces used in this monograph are always defined on a bounded (open, connected) domain $\Omega \subset \mathbb{R}^n$ or on a subset $\Gamma \subseteq \partial\Omega$ of the boundary $\partial\Omega$. We always assume that $\partial\Omega$, Γ are at least Lipschitz continuous which includes in particular convex polyhedral domains Ω.

We will be making frequent use of a *fictitious domain* $\square$, preferably a cube like $(0,1)^n$, which contains Ω,

$$\Omega \subseteq \square. \tag{3.1.1}$$

Furthermore, we adopt the standard notation of fractional Sobolev spaces $H^s(\Omega)$ and $H^s(\square)$ for $s \in \mathbb{R}$ from e.g. [Gr, LM] endowed with their usual norms. Sobolev spaces on $\partial\Omega$ can be viewed as trace spaces by using factor norms

$$\|h\|_{H^s(\partial\Omega)} = \inf_{f \in H^{s+1/2}(\Omega),\ f|_{\partial\Omega} = h} \|f\|_{H^{s+1/2}(\Omega)}$$

which are for a certain range of s (except for s equal to an integer) equivalent to the definition via an atlas and a partition of unity, see e.g. [Gr], Section 1.3, for the definition. We always assume to work in this range. When Ω is a Lipschitz domain, this is the case for $0 < s < 1$. Correspondingly, the range for s increases when Ω is a Hölder continuous domain of higher smoothness. Equivalent norms for the same space will not be distinguished.

One situation where the smoothness of Ω comes into play are the constants in the classical Trace Theorem, Theorem 4.9 below.

3.2 Multiscale Decomposition of Function Spaces — Uniform Refinements

Since solutions of elliptic boundary value problems by means of weak formulations usually belong to Sobolev spaces, a decomposition of such spaces on domains and boundary manifolds will play a substantial role in the analysis. The situation will be as follows.

Let H^s, $s \in \mathbb{R}$, be a scale of Sobolev spaces such that

$$H^s \subseteq L_2 \quad \text{for } s \geq 0 \tag{3.2.1}$$

holds, and for negative s the space H^s is the dual of $(H^{-s})'$. The norms and inner products corresponding to H^s etc. are indexed accordingly. In view of Section 3.1, L_2 will either play the role of $L_2(\square)$ or $L_2(\Gamma)$, and correspondingly H^s. We will always denote by n the dimension of the underlying physical domain Ω.

In order to apply the techniques from Chapter 2, a scale (3.2.1) is considered for each of the spaces $H_{i,0}$ in (2.3), $i = 1, \ldots, M$. That is, for some $s = s(i)$ one has $H_{i,0} = H^s$. Since the construction of wavelets recalled in this chapter is independent of the other spaces, we focus here on the construction for one such scale H^s and skip the index i to simplify the notation.

3.2.1 Multiresolution of L_2

Practical constructions of wavelets typically start out with multiresolution analyses of function spaces. In the following, a number of corresponding necessary notions and results which do not rely on Fourier techniques are collected from the survey article [D3]. An example of the situation described in this subsection is given in Section 3.2.5 below.

Starting with a fixed parameter $j_0 \in \mathbb{N}_0$, consider a *multiresolution* $\mathcal{S}$ of L_2 which consists of closed subspaces S_j of L_2, called *trial spaces*, such that they are nested and their union is dense in L_2,

$$S_{j_0} \subset S_{j_0+1} \subset \ldots \subset S_j \subset S_{j+1} \subset \ldots L_2, \qquad \operatorname{clos}_{L_2}\left(\bigcup_{j=j_0}^{\infty} S_j\right) = L_2. \tag{3.2.2}$$

The index j stands for the *level of resolution* or *refinement level*, and j_0 always denotes the *coarsest* level. Recall that for any finite subset $\Theta \subset L_2$ the linear span of Θ is abbreviated as

$$S(\Theta) = \operatorname{span}\{\Theta\}.$$

Typically the multiresolution spaces S_j have the form

$$S_j = S(\Phi_j), \qquad \Phi_j = \{\phi_{j,k} : k \in \Delta_j\}, \tag{3.2.3}$$

for some finite index set Δ_j, where the set $\{\Phi_j\}_{j=j_0}^{\infty}$ is *uniformly stable* in the sense that

$$\|\mathbf{c}\|_{\ell_2(\Delta_j)} \sim \|\mathbf{c}^T\Phi_j\|_{L_2}, \qquad \mathbf{c} = \{c_k\}_{k\in\Delta_j} \in \ell_2(\Delta_j), \tag{3.2.4}$$

holds uniformly in j. Here we have used again the shorthand notation

$$\mathbf{c}^T\Phi_j = \sum_{k\in\Delta_j} c_k\phi_{j,k}$$

introduced in Chapter 2. Also Φ_j denotes both the (column) vector containing the functions $\phi_{j,k}$ as well as the set of functions (3.2.3).

The collection Φ_j is called *single scale basis* since all its members live only on one scale j. In the present context of multiresolution analysis, it is also called *generator basis* or shortly *generators* (of the multiresolution). We assume that the $\phi_{j,k}$ are compactly supported with

$$\operatorname{diam}(\operatorname{supp}\phi_{j,k}) \sim 2^{-j}. \tag{3.2.5}$$

It follows from (3.2.4) that they are scaled such that

$$\|\phi_{j,k}\|_{L_2} \sim 1 \tag{3.2.6}$$

holds. It is known that nestedness (3.2.2) together with stability (3.2.4) implies the existence of matrices $\mathbf{M}_{j,0} = (m^j_{r,k})_{r\in\Delta_{j+1},k\in\Delta_j}$ such that the two-scale relation

$$\phi_{j,k} = \sum_{r\in\Delta_{j+1}} m^j_{r,k}\phi_{j+1,r}, \quad k\in\Delta_j, \tag{3.2.7}$$

is satisfied. The presentation of the material will be essentially simplified by viewing (3.2.7) as a matrix–vector equation which then takes on the compact form

$$\Phi_j = \mathbf{M}^T_{j,0}\Phi_{j+1}. \tag{3.2.8}$$

Any set of functions satisfying an equation of this form, the *refinement* or *two-scale relation,* will be called *refinable.*

Denoting by $[X, Y]$ the space of bounded linear operators from a normed linear space X into the normed linear space Y, one has that

$$\mathbf{M}_{j,0} \in [\ell_2(\Delta_j), \ell_2(\Delta_{j+1})]$$

is *uniformly sparse* which means that the number of entries in each row or column is uniformly bounded. Furthermore, one infers from (3.2.4) that

$$\|\mathbf{M}_{j,0}\| = \mathcal{O}(1), \quad j\ge j_0, \tag{3.2.9}$$

where the corresponding operator norm is defined as

$$\|\mathbf{M}_{j,0}\| := \sup_{\mathbf{c}\in\ell_2(\Delta_j),\ \|\mathbf{c}\|_{\ell_2(\Delta_j)}=1} \|\mathbf{M}_{j,0}\mathbf{c}\|_{\ell_2(\Delta_{j+1})}.$$

Since the union of $\mathcal{S}$ is dense in L_2, a basis for L_2 can be assembled from functions which span any complement between two successive spaces S_j and S_{j+1}, i.e.,

$$S(\Phi_{j+1}) = S(\Phi_j) \oplus S(\Psi_j) \tag{3.2.10}$$

where

$$\Psi_j = \{\psi_{j,k} : k \in \nabla_j\}, \qquad \nabla_j := \Delta_{j+1} \setminus \Delta_j. \tag{3.2.11}$$

The functions Ψ_j are called *wavelet functions* or shortly *wavelets* if, among other conditions detailed below, the union $\{\Phi_j \cup \Psi_j\}$ is still uniformly stable in the sense of (3.2.4). Since (3.2.10) implies $S(\Psi_j) \subset S(\Phi_{j+1})$, the functions in Ψ_j must also satisfy a matrix–vector relation of the form

$$\Psi_j = \mathbf{M}_{j,1}^T \Phi_{j+1} \tag{3.2.12}$$

with a matrix $\mathbf{M}_{j,1}$ of size $(\#\Delta_{j+1}) \times (\#\nabla_j)$. Furthermore, (3.2.10) is equivalent to the fact that the linear operator composed of $\mathbf{M}_{j,0}$ and $\mathbf{M}_{j,1}$,

$$\mathbf{M}_j = (\mathbf{M}_{j,0}, \mathbf{M}_{j,1}), \tag{3.2.13}$$

is *invertible* as a mapping from $\ell_2(\Delta_j \cup \nabla_j)$ onto $\ell_2(\Delta_{j+1})$. One can also show that $\{\Phi_j \cup \Psi_j\}$ is uniformly stable if and only if

$$\|\mathbf{M}_j\|, \|\mathbf{M}_j^{-1}\| = \mathcal{O}(1), \quad j \to \infty. \tag{3.2.14}$$

The particular cases that will be important for practical purposes are when not only $\mathbf{M}_{j,0}$ and $\mathbf{M}_{j,1}$ are uniformly sparse but also the inverse of $\mathbf{M}_j$ denoted by $\mathbf{G}_j$, which is assumed to be blocked into

$$\mathbf{G}_j = \mathbf{M}_j^{-1} = \begin{pmatrix} \mathbf{G}_{j,0} \\ \mathbf{G}_{j,1} \end{pmatrix}. \tag{3.2.15}$$

A special situation occurs when

$$\mathbf{G}_j = \mathbf{M}_j^{-1} = \mathbf{M}_j^{-T}$$

which corresponds to the case of (L_2) *orthogonal wavelets* [Dau]. Examples for general $\mathbf{M}_j$, $\mathbf{G}_j$ can be found in [D3, DKU2].

Thus, the identification of the functions Ψ_j which span the complement of $S(\Phi_j)$ in $S(\Phi_{j+1})$ is equivalent to completing a given refinement matrix $\mathbf{M}_{j,0}$ to a 'square' matrix $\mathbf{M}_j$ in such a way that (3.2.14) is satisfied. Any such completion $\mathbf{M}_{j,1}$ is called *stable completion* of $\mathbf{M}_{j,0}$. In other words, the problem of the construction of compactly supported wavelets can equivalently be formulated as an algebraic problem of finding the (uniformly) sparse completion of a (uniformly) sparse matrix $\mathbf{M}_{j,0}$ in such a way that its inverse is also (uniformly) sparse. The fact that inverses of sparse matrices are usually dense elucidates the difficulties in the constructions. The concept of stable completions has been introduced in [CDP]. A special case known by now as the *lifting scheme* is discussed in [Sw]. Of course, constructions that yield compactly supported wavelets are particularly suited for computations in numerical analysis.

Combining the two–scale relations (3.2.8) and (3.2.12), one can see that $\mathbf{M}_j$ performs a change of bases in the space S_{j+1},

$$\begin{pmatrix} \Phi_j \\ \Psi_j \end{pmatrix} = \begin{pmatrix} \mathbf{M}_{j,0}^T \\ \mathbf{M}_{j,1}^T \end{pmatrix} \Phi_{j+1} = \mathbf{M}_j^T \Phi_{j+1}. \tag{3.2.16}$$

Conversely, applying the inverse of $\mathbf{M}_j$ to both sides of (3.2.16) results in the *reconstruction identity*

$$\Phi_{j+1} = \mathbf{G}_j^T \begin{pmatrix} \Phi_j \\ \Psi_j \end{pmatrix} = \mathbf{G}_{j,0}^T \Phi_j + \mathbf{G}_{j,1}^T \Psi_j. \tag{3.2.17}$$

Fixing a *finest resolution level* J, one can repeat the decomposition (3.2.10) so that $S_J = S(\Phi_J)$ can be written in terms of the functions from the coarsest space supplied with the complement functions from all intermediate levels,

$$S(\Phi_J) = S(\Phi_{j_0}) \oplus \bigoplus_{j=j_0}^{J-1} S(\Psi_j). \tag{3.2.18}$$

Thus, every function $v \in S(\Phi_J)$ can be written in its *single–scale representation*

$$v = (\mathbf{c}_J)^T \Phi_J = \sum_{k \in \Delta_J} c_{J,k} \phi_{J,k} \tag{3.2.19}$$

as well as in its *multi–scale form*

$$v = (\mathbf{c}_{j_0})^T \Phi_{j_0} + (\mathbf{d}_{j_0})^T \Psi_{j_0} + \cdots + (\mathbf{d}_{J-1})^T \Psi_{J-1} \tag{3.2.20}$$

with respect to the *multiscale* or *wavelet basis*

$$\Psi^J := \Phi_{j_0} \cup \bigcup_{j=j_0}^{J-1} \Psi_j =: \bigcup_{j=j_0-1}^{J-1} \Psi_j \tag{3.2.21}$$

Often the single–scale representation of a function may be easier to compute and evaluate while the multi–scale representation allows one to separate features of the underlying function characterized by different length scales. Since therefore both representations are advantageous, it is useful to determine the transformation between the two representations, commonly referred to as the *Wavelet Transform*,

$$\mathbf{T}_J : \ell_2(\Delta_j) \to \ell_2(\Delta_j), \qquad \mathbf{d}^J \mapsto \mathbf{c}_J, \tag{3.2.22}$$

where

$$\mathbf{d}^J := (\mathbf{c}_{j_0}, \mathbf{d}_{j_0}, \ldots, \mathbf{d}_{J-1})^T.$$

The previous relations (3.2.16) and (3.2.17) indicate that this will involve the matrices $\mathbf{M}_j$ and $\mathbf{G}_j$. In fact, $\mathbf{T}_J$ has the representation

$$\mathbf{T}_J = \mathbf{T}_{J,J-1} \cdots \mathbf{T}_{J,j_0}, \tag{3.2.23}$$

where each factor has the form

$$\mathbf{T}_{J,j} := \begin{pmatrix} \mathbf{M}_j & \mathbf{0} \\ \mathbf{0} & \mathbf{I}^{(\#\Delta_J - \#\Delta_j)} \end{pmatrix} \in \mathbb{R}^{(\#\Delta_J)\times(\#\Delta_J)}. \tag{3.2.24}$$

Schematically $\mathbf{T}_J$ can be visualized as a pyramid scheme,

$$\begin{array}{ccccccccc} & \mathbf{M}_{j_0,0} & & \mathbf{M}_{j_0+1,0} & & & & \mathbf{M}_{J-1,0} & \\ \mathbf{c}_{j_0} & \longrightarrow & \mathbf{c}_{j_0+1} & \longrightarrow & \mathbf{c}_{j_0+2} & \longrightarrow \cdots & \mathbf{c}_{J-1} & \longrightarrow & \mathbf{c}_J \\ & \mathbf{M}_{j_0,1} & & \mathbf{M}_{j_0+1,1} & & & & \mathbf{M}_{J-1,1} & \\ & \nearrow & & \nearrow & & \nearrow \cdots & & \nearrow & \\ \mathbf{d}_{j_0} & & \mathbf{d}_{j_0+1} & & \mathbf{d}_{j_0+2} & & \mathbf{d}_{J-1} & & \end{array} \tag{3.2.25}$$

Accordingly, the inverse transform $\mathbf{T}_J^{-1}$ can be written also in product structure (3.2.23) in reverse order involving the matrices $\mathbf{G}_j$ as follows:

$$\mathbf{T}_J^{-1} = \mathbf{T}_{J,j_0}^{-1} \cdots \mathbf{T}_{J,J-1}^{-1}, \tag{3.2.26}$$

where each factor has the form

$$\mathbf{T}_{J,j}^{-1} := \begin{pmatrix} \mathbf{G}_j & \mathbf{0} \\ \mathbf{0} & \mathbf{I}^{(\#\Delta_J - \#\Delta_j)} \end{pmatrix} \in \mathbb{R}^{(\#\Delta_J)\times(\#\Delta_J)}. \tag{3.2.27}$$

The corresponding pyramid scheme is then

$$\begin{array}{ccccccccc} & \mathbf{G}_{J-1,0} & & \mathbf{G}_{J-2,0} & & & & \mathbf{G}_{j_0,0} & \\ \mathbf{c}_J & \longrightarrow & \mathbf{c}_{J-1} & \longrightarrow & \mathbf{c}_{J-2} & \longrightarrow & \cdots & \longrightarrow & \mathbf{c}_{j_0} \\ & \mathbf{G}_{J-1,1} & & \mathbf{G}_{J-2,1} & & & & \mathbf{G}_{j_0,1} & \\ & \searrow & & \searrow & & \searrow & \cdots & \searrow & \\ & & \mathbf{d}_{J-1} & & \mathbf{d}_{J-2} & & \mathbf{d}_{J-1} & & \mathbf{d}_{j_0} \end{array} \tag{3.2.28}$$

Remark 3.1 *Property (3.2.14) and the fact that $\mathbf{M}_j$ and $\mathbf{G}_j$ can be applied in $(\#\Delta_{j+1})$ operations uniformly in j entails that the complexity of applying $\mathbf{T}_J$ or $\mathbf{T}_J^{-1}$ using the pyramid scheme is of order $\mathcal{O}(\#\Delta_J) = \mathcal{O}(\dim S(\Phi_J))$ uniformly in J. This justifies the expression* Fast Wavelet Transform. *Note that there is no need to explicitly assemble $\mathbf{T}_J$ or $\mathbf{T}_J^{-1}$.*

We will see later in Section 3.2.4 how the transformation $\mathbf{T}_J$ may be used for preconditioning.

Since $\mathcal{S}$ is dense in L_2, a basis for the whole space L_2 may be given by letting $J \to \infty$ in (3.2.21),

$$\Psi := \Phi_{j_0} \cup \bigcup_{j=j_0}^{\infty} \Psi_j = \{\psi_{j,k} : (j,k) \in \mathbb{I}\}, \qquad \mathbb{I} := \{\{j_0\} \times \Delta_{j_0}\} \cup \bigcup_{j=j_0}^{\infty} \{\{j\} \times \nabla_j\}. \tag{3.2.29}$$

The next theorem from [D1, D2] illustrates the interdependence between Ψ and $\mathbf{T}_J$.

Theorem 3.2 *The multiscale transformations $\mathbf{T}_J$ are well-conditioned in the sense*

$$\|\mathbf{T}_J\|, \|\mathbf{T}_J^{-1}\| = \mathcal{O}(1), \quad J \geq j_0, \tag{3.2.30}$$

if and only if the collection Ψ defined by (3.2.29) is a Riesz basis *for L_2, i.e., every $v \in L_2$ has unique expansions*

$$v = \sum_{j=j_0-1}^{\infty} \langle v, \tilde{\Psi}_j \rangle \Psi_j = \sum_{j=j_0-1}^{\infty} \langle v, \Psi_j \rangle \tilde{\Psi}_j, \tag{3.2.31}$$

where $\tilde{\Psi}$ defined analogously as in (3.2.29) is also a Riesz basis for L_2 which is biorthogonal *or* dual *to Ψ,*

$$\langle \Psi, \tilde{\Psi} \rangle = \mathbf{I} \tag{3.2.32}$$

such that

$$\|v\|_{L_2} \sim \|\langle v, \tilde{\Psi} \rangle^T\|_{\ell_2(\mathbb{I})} \sim \|\langle v, \Psi \rangle^T\|_{\ell_2(\mathbb{I})}. \tag{3.2.33}$$

We briefly explain next how the functions in $\tilde{\Psi}$, denoted as *wavelets dual to* Ψ, or *dual wavelets*, can be determined.

To this end, assume that there is a second multiresolution $\tilde{\mathcal{S}}$ of L_2 satisfying (3.2.2) where

$$\tilde{S}_j = S(\tilde{\Phi}_j), \qquad \tilde{\Phi}_j = \{\tilde{\phi}_{j,k} : k \in \Delta_j\} \tag{3.2.34}$$

and $\{\tilde{\Phi}_j\}_{j=j_0}^{\infty}$ is uniformly stable in j in the sense of (3.2.4). Let the functions in $\tilde{\Phi}_j$ also have compact support satisfying (3.2.5). Furthermore, suppose that the biorthogonality conditions

$$\langle \Phi_j, \tilde{\Phi}_j \rangle = \mathbf{I} \tag{3.2.35}$$

hold. We will often refer to Φ_j as the *primal* and to $\tilde{\Phi}_j$ as the *dual generators*. The nestedness of the $\tilde{S}_j$ and the stability again implies that $\tilde{\Phi}_j$ is refinable with some matrix $\tilde{\mathbf{M}}_{j,0}$, similar to (3.2.8),

$$\tilde{\Phi}_j = \tilde{\mathbf{M}}_{j,0}^T \tilde{\Phi}_{j+1}. \tag{3.2.36}$$

The problem of determining the biorthogonal wavelets now consists in finding bases Ψ_j, $\tilde{\Psi}_j$ for the complements of $S(\Phi_j)$ in $S(\Phi_{j+1})$, and of $S(\tilde{\Phi}_j)$ in $S(\tilde{\Phi}_{j+1})$, such that

$$S(\Phi_j) \perp S(\tilde{\Psi}_j), \qquad S(\tilde{\Phi}_j) \perp S(\Psi_j) \tag{3.2.37}$$

and

$$S(\Psi_j) \perp S(\tilde{\Psi}_r), \quad j \neq r, \tag{3.2.38}$$

holds. The connection between the concept of stable completions and the dual generators and wavelets is made by the following result which is a special case from [CDP].

Proposition 3.3 *Suppose that the biorthogonal collections* $\{\Phi_j\}_{j=j_0}^{\infty}$, $\{\tilde{\Phi}_j\}_{j=j_0}^{\infty}$ *are both uniformly stable and refinable with refinement matrices* $\mathbf{M}_{j,0}$, $\tilde{\mathbf{M}}_{j,0}$, *i.e.,*

$$\Phi_j = \mathbf{M}_{j,0}^T \Phi_{j+1}, \qquad \tilde{\Phi}_j = \tilde{\mathbf{M}}_{j,0}^T \tilde{\Phi}_{j+1}, \tag{3.2.39}$$

and satisfy the duality condition (3.2.35). Assume that $\check{\mathbf{M}}_{j,1}$ *is any stable completion of* $\mathbf{M}_{j,0}$ *such that*

$$\check{\mathbf{M}}_j := (\mathbf{M}_{j,0}, \check{\mathbf{M}}_{j,1}) = \check{\mathbf{G}}_j^{-1} \tag{3.2.40}$$

satisfies (3.2.14).
Then

$$\mathbf{M}_{j,1} := (\mathbf{I} - \mathbf{M}_{j,0} \tilde{\mathbf{M}}_{j,0}^T) \check{\mathbf{M}}_{j,1} \tag{3.2.41}$$

is also a stable completion of $\mathbf{M}_{j,0}$, *and* $\mathbf{G}_j = \mathbf{M}_j^{-1} = (\mathbf{M}_{j,0}, \mathbf{M}_{j,1})^{-1}$ *has the form*

$$\mathbf{G}_j = \begin{pmatrix} \tilde{\mathbf{M}}_{j,0}^T \\ \check{\mathbf{G}}_{j,1} \end{pmatrix}. \tag{3.2.42}$$

Moreover, the collections of functions

$$\Psi_j := \mathbf{M}_{j,1}^T \Phi_{j+1}, \qquad \tilde{\Psi}_j := \check{\mathbf{G}}_{j,1} \tilde{\Phi}_{j+1} \tag{3.2.43}$$

form biorthogonal systems,

$$\langle \Psi_j, \tilde{\Psi}_j \rangle = \mathbf{I}, \qquad \langle \Psi_j, \tilde{\Phi}_j \rangle = \langle \Phi_j, \tilde{\Psi}_j \rangle = \mathbf{0}, \tag{3.2.44}$$

so that

$$S(\Psi_j) \perp S(\tilde{\Psi}_r), \quad j \neq r, \qquad S(\Phi_j) \perp S(\tilde{\Psi}_j), \quad S(\tilde{\Phi}_j) \perp S(\Psi_j). \tag{3.2.45}$$

In particular, the relations (3.2.35), (3.2.44) imply that the collections

$$\Psi = \bigcup_{j=j_0-1}^{\infty} \Psi_j, \qquad \tilde{\Psi} := \bigcup_{j=j_0-1}^{\infty} \tilde{\Psi}_j := \tilde{\Phi}_{j_0} \cup \bigcup_{j=j_0}^{\infty} \tilde{\Psi}_j \tag{3.2.46}$$

are biorthogonal,

$$\langle \Psi, \tilde{\Psi} \rangle = \mathbf{I}. \tag{3.2.47}$$

Remark 3.4 *It is important to note that the properties needed in addition to (3.2.47) in order to ensure (3.2.33) are neither properties of the complements nor of their bases* Ψ, $\tilde{\Psi}$ *but of the multiresolution sequences* $\mathcal{S}$ *and* $\tilde{\mathcal{S}}$. *These can be phrased as approximation and regularity properties and will be described in the next subsection.*

We briefly recall yet another useful point of view. The operators

$$\begin{aligned} P_j v &:= \langle v, \tilde{\Phi}_j \rangle \Phi_j = \langle v, \tilde{\Psi}^j \rangle \Psi^j = \langle v, \tilde{\Phi}_{j_0} \rangle \Phi_{j_0} + \sum_{r=j_0}^{j-1} \langle v, \tilde{\Psi}_r \rangle \Psi_r \\ P'_j v &:= \langle v, \Phi_j \rangle \tilde{\Phi}_j = \langle v, \Psi^j \rangle \tilde{\Psi}^j = \langle v, \Phi_{j_0} \rangle \tilde{\Phi}_{j_0} + \sum_{r=j_0}^{j-1} \langle v, \Psi_r \rangle \tilde{\Psi}_r \end{aligned} \tag{3.2.48}$$

are projectors onto

$$S(\Phi_j) = S(\Psi^j) \quad \text{and} \quad S(\tilde{\Phi}_j) = S(\tilde{\Psi}^j) \tag{3.2.49}$$

respectively, which satisfy

$$P_r P_j = P_r, \quad P'_r P'_j = P'_r, \qquad r \le j. \tag{3.2.50}$$

Remark 3.5 *Let* $\{\Phi_j\}_{j=j_0}^{\infty}$ *be uniformly stable. The* P_j *defined by (3.2.48) are uniformly bounded if and only if* $\{\tilde{\Phi}_j\}_{j=j_0}^{\infty}$ *is also uniformly stable. Moreover, the* P_j *satisfy (3.2.50) if and only if the* $\tilde{\Phi}_j$ *are refinable as well. Note that then (3.2.35) implies*

$$\mathbf{M}_{j,0}^T \tilde{\mathbf{M}}_{j,0} = \mathbf{I}. \tag{3.2.51}$$

In terms of the projectors, the uniform stability of the complement bases Ψ_j, $\tilde{\Psi}_j$ means that

$$\|(P_{j+1} - P_j)v\|_{L_2} \sim \|\langle v, \tilde{\Psi}_j \rangle^T\|_{\ell_2(\nabla_j)}, \quad \|(P'_{j+1} - P'_j)v\|_{L_2} \sim \|\langle v, \Psi_j \rangle^T\|_{\ell_2(\nabla_j)}, \tag{3.2.52}$$

so that the L_2 norm equivalence (3.2.33) is equivalent to

$$\|v\|_{L_2}^2 \sim \sum_{j=j_0}^{\infty} \|(P_j - P_{j-1})v\|_{L_2}^2 \sim \sum_{j=j_0}^{\infty} \|(P'_j - P'_{j-1})v\|_{L_2}^2 \tag{3.2.53}$$

for any $v \in L_2$, where $P_{j_0-1} = P'_{j_0-1} := 0$.

The whole concept that has been derived so far lives from both Φ_j *and* $\tilde{\Phi}_j$. It should be pointed out that in the algorithms one actually does not need $\tilde{\Phi}_j$ for computations. Indeed, in some cases like the construction of finite element based wavelets on non-uniform meshes [DSt] it is not so easy to explicitly construct $\tilde{\Phi}_j$ and $\tilde{\Psi}_j$. Nevertheless, for establishing (3.2.53) or the results in the next Section 3.2.2, it is sufficient to replace the duality conditions (3.2.35) by a reverse Cauchy–Schwarz inequality

$$\inf_{v \in S(\Phi_j)} \sup_{\tilde{v} \in S(\tilde{\Phi}_j)} \frac{\langle v, \tilde{v} \rangle}{\|v\|_{L_2} \|\tilde{v}\|_{L_2}} \geq c_{\text{stab}} \tag{3.2.54}$$

uniformly in j. By the same arguments as in the proof of Lemma 3.12 below, one can show that (3.2.35) implies (3.2.54).

We recall next results that guarantee norm equivalences of the type (2.18) for Sobolev spaces.

3.2.2 Multiresolution of Sobolev Spaces and Norm Equivalences

Let now $\mathcal{S}$ be a multiresolution sequence consisting of closed subspaces of H^s with the property (3.2.2) whose union is dense in H^s. The following result from [D2] ensures under which conditions norm equivalences hold for the H^s–norm.

Theorem 3.6 *Let* $\{\Phi_j\}_{j=j_0}^{\infty}, \{\tilde{\Phi}_j\}_{j=j_0}^{\infty}$ *be uniformly stable, refinable, biorthogonal collections and let the* $P_j : H^s \to S(\Phi_j)$ *be defined by (3.2.48).*
If the Jackson-type estimate

$$\inf_{v_j \in S_j} \|v - v_j\|_{L_2} \lesssim 2^{-sj} \|v\|_{H^s}, \quad v \in H^s, \ 0 < s \leq \bar{d}, \tag{3.2.55}$$

and the Bernstein inequality

$$\|v_j\|_{H^s} \lesssim 2^{sj} \|v_j\|_{L_2}, \quad v_j \in S_j, \ s < \bar{t}, \tag{3.2.56}$$

hold for

$$S_j = \left\{ \begin{array}{c} S(\Phi_j) \\ S(\tilde{\Phi}_j) \end{array} \right\} \ \textit{with}\ \text{order}\ \bar{d} = \left\{ \begin{array}{c} d \\ \tilde{d} \end{array} \right\} \ \textit{and}\ \bar{t} = \left\{ \begin{array}{c} t \\ \tilde{t} \end{array} \right\}, \tag{3.2.57}$$

then for

$$0 < \sigma := \min\{d, t\}, \qquad 0 < \tilde{\sigma} := \min\{\tilde{d}, \tilde{t}\}, \tag{3.2.58}$$

one has

$$\|v\|_{H^s}^2 \sim \sum_{j=j_0}^{\infty} 2^{2sj} \|(P_j - P_{j-1})v\|_{L_2}^2, \quad s \in (-\tilde{\sigma}, \sigma). \tag{3.2.59}$$

Recall here that $H^s = (H^{-s})'$ for $s < 0$.

The regularity of $\mathcal{S}$ and $\tilde{\mathcal{S}}$ is characterized by

$$t := \sup\{s : S(\Phi_j) \subset H^s, \ j \geq j_0\}, \qquad \tilde{t} := \sup\{s : S(\tilde{\Phi}_j) \subset H^s, \ j \geq j_0\} \tag{3.2.60}$$

Recalling the representation (3.2.52), one can immediately derive the following fact.

Corollary 3.7 *Suppose that the assumptions in Theorem 3.6 hold. Then the norm equivalence*

$$\|v\|_{H^s}^2 \sim \sum_{j=j_0-1}^{\infty} 2^{2sj} \|\langle v, \tilde{\Psi}_j\rangle^T\|_{\ell_2(\nabla_j)}^2, \quad s \in (-\tilde{\sigma}, \sigma). \tag{3.2.61}$$

is valid.

Note that, in particular for $s = 0$ the Riesz basis property relative to L_2 of the Ψ, $\tilde{\Psi}$ (3.2.33) is recovered.

In many applications one needs (3.2.59) or (3.2.61) only for certain $s > 0$. Then it suffices to require (3.2.55) and (3.2.56) for $\{\Phi_j\}_{j=j_0}^{\infty}$.

Theorem 3.6 can also be interpreted as follows. The mapping

$$\Sigma_s v := \sum_{j=j_0}^{\infty} 2^{sj}(P_j - P_{j-1})v \tag{3.2.62}$$

acts like a Bessel potential operator as a *shift* in the scale H^s, i.e.,

$$\|\Sigma_s v\|_{H^t} \sim \|v\|_{H^{s+t}}, \quad s + t \in (-\tilde{\sigma}, \sigma). \tag{3.2.63}$$

The Jackson estimates (3.2.55) of order $\tilde{d}$ for $S(\tilde{\Phi}_j)$ imply the *cancellation properties*

$$|\langle v, \psi_{j,k}\rangle| \lesssim 2^{-j(n/2+\tilde{d})} \|v\|_{W_\infty^{\tilde{d}}(\mathrm{supp}\,\psi_{j,k})}, \tag{3.2.64}$$

where $W_\infty^{\tilde{d}} \subset L_\infty$ is the Sobolev space of order $\tilde{d}$.

Remark 3.8 *When the wavelets live on $\Omega \subset \mathbb{R}^n$, (3.2.55) means that all polynomials up to order $\tilde{d}$ are contained in $S(\tilde{\Phi}_j)$. One also says that $S(\tilde{\Phi}_j)$ is* exact *of order $\tilde{d}$. On account of (3.2.32), this implies that the wavelets $\psi_{j,k}$ are orthogonal to polynomials up to order $\tilde{d}$ or have $\tilde{d}$th order* vanishing moments. *By Taylor expansion, this in turn yields (3.2.64).*

We will later be using the following generalization of the discrete norms (3.2.53). Let for $s \in \mathbb{R}$

$$|\!|\!| v |\!|\!|_s := \left(\sum_{j=j_0}^{\infty} 2^{2sj} \|(P_j - P_{j-1})v\|_{L_2}^2 \right)^{1/2} \tag{3.2.65}$$

which by the relations (3.2.52) is also equivalent to

$$|v|_s := \left(\sum_{j=j_0-1}^{\infty} 2^{2sj} \|\langle v, \tilde{\Psi}_j\rangle^T\|_{\ell_2(\nabla_j)}^2 \right)^{1/2}. \tag{3.2.66}$$

In this notation, (3.2.59) and (3.2.61) read

$$\|v\|_{H^s} \sim |\!|\!| v |\!|\!|_s \sim |v|_s. \tag{3.2.67}$$

In terms of such discrete norms, Jackson and Bernstein estimates hold with constants equal to 1, which turns out to be very useful later in Chapter 4.

Lemma 3.9 *Let $\{\Phi_j\}_{j=j_0}^{\infty}, \{\tilde{\Phi}_j\}_{j=j_0}^{\infty}$ be uniformly stable, refinable, biorthogonal collections and let the P_j be defined by (3.2.48). Then the estimates*

$$\|v - P_j v\|_{s'} \leq 2^{-(j+1)(s-s')} \|v\|_s, \qquad v \in H^s, \quad s' \leq s \leq d, \tag{3.2.68}$$

and

$$\|v_j\|_s \leq 2^{j(s-s')} \|v_j\|_{s'}, \qquad v_j \in S(\Phi_j), \quad s' \leq s \leq d, \tag{3.2.69}$$

are valid, and correspondingly for the dual side.

Proof: The easy proof is included for completeness. Let $v \in H^s$. By definition (3.2.65), one has

$$\begin{aligned}\|v - P_j v\|_{s'}^2 &= \sum_{r=j+1}^{\infty} 2^{2s'r} \|(P_r - P_{r-1})v\|_{L_2}^2 = \sum_{r=j+1}^{\infty} 2^{-2(s-s')r} 2^{2sr} \|(P_r - P_{r-1})v\|_{L_2}^2 \\ &\leq 2^{-2(j+1)(s-s')} \|v\|_s^2\end{aligned}$$

for $s' \leq s$. Similarly, one can estimate for any $v_j \in S(\Phi_j)$

$$\begin{aligned}\|v_j\|_s^2 &= \sum_{r=j_0}^{j} 2^{2sr} \|(P_r - P_{r-1})v\|_{L_2}^2 = \sum_{r=j_0}^{j} 2^{2(s-s')r} 2^{2rs'} \|(P_r - P_{r-1})v\|_{L_2}^2 \\ &\leq 2^{2j(s-s')} \|v_j\|_{s'}^2\end{aligned}$$

for $s \geq s'$. ∎

The same reasoning applies to the norm $|\cdot|$ defined in (3.2.66).

Lemma 3.10 *Let $\{\Phi_j\}_{j=j_0}^{\infty}, \{\tilde{\Phi}_j\}_{j=j_0}^{\infty}$ be uniformly stable, refinable, biorthogonal collections and let the P_j be defined by (3.2.48). Then the estimates*

$$|v - P_j v|_{s'} \leq 2^{-(j+1)(s-s')} |v|_s, \qquad v \in H^s, \quad s' \leq s \leq d, \tag{3.2.70}$$

and

$$|v_j|_s \leq 2^{j(s-s')} |v_j|_{s'}, \qquad v_j \in S(\Phi_j), \quad s' \leq s \leq d, \tag{3.2.71}$$

hold, and correspondingly for the dual side.

3.2.3 Reverse Cauchy–Schwarz Inequalities

The biorthogonality condition (3.2.35) or its generalized variant (3.2.54) implies together with direct and inverse estimates the following reverse Cauchy–Schwarz inequalities for finite-dimensional spaces [DK2]. It will be one essential ingredient for the discussion of the LBB condition in Section 4.2.4.

Lemma 3.11 *Let the assumptions in Theorem 3.6 be valid such that the norm equivalence (3.2.59) holds for $(-\tilde{\sigma}, \sigma)$ with $\sigma, \tilde{\sigma}$ defined in (3.2.58). Then for any $v \in S(\Phi_j)$ there exists some $\tilde{v}^* = \tilde{v}^*(v) \in S(\tilde{\Phi}_j)$ such that*

$$\|v\|_{H^s} \|\tilde{v}^*\|_{H^{-s}} \lesssim \langle v, \tilde{v}^* \rangle \tag{3.2.72}$$

for any $0 \leq s < \min(\sigma, \tilde{\sigma})$.

The proof of this result given in [DK2] for $s = 1/2$ in terms of the projectors P_j defined in (3.2.48) and corresponding duals P_j' immediately carries over to more general s. Recalling the representation (3.2.49) in terms of wavelets, the reverse Cauchy inequality (3.2.72) attains the following sharp form.

Lemma 3.12 *Let the assumptions of Lemma 3.10 hold. Then for every $v \in S(\Phi_j)$ there exists some $\tilde{v}^* = \tilde{v}^*(v) \in S(\tilde{\Phi}_j)$ such that*

$$|v|_s \, |\tilde{v}^*|_{-s} = \langle v, \tilde{v}^* \rangle \tag{3.2.73}$$

for any $0 \le s \le \min(\sigma, \tilde{\sigma})$.

Proof: Every $v \in S(\Phi_j)$ can be written as

$$v = \sum_{r=j_0-1}^{j-1} 2^{sr} \sum_{k \in \nabla_r} v_{r,k} \psi_{r,k}.$$

Setting now

$$\tilde{v}^* := \sum_{r=j_0-1}^{j-1} 2^{-sr} \sum_{k \in \nabla_r} v_{r,k} \tilde{\psi}_{r,k}$$

with the same coefficients $v_{j,k}$, the definition of $|\cdot|_s$ yields by biorthogonality (3.2.47)

$$|v|_s \, |\tilde{v}^*|_{-s} = \sum_{r=j_0-1}^{j-1} \sum_{k \in \nabla_r} |v_{j,k}|^2.$$

Combining this with the observation

$$\langle v, \tilde{v}^* \rangle = \sum_{r=j_0-1}^{j-1} \sum_{k \in \nabla_r} |v_{j,k}|^2$$

confirms (3.2.73). ∎

Remark 3.13 *The previous proof reveals that the identity (3.2.73) is also true for elements from infinite-dimensional spaces H^s and $(H^s)'$ for which Ψ and $\tilde{\Psi}$ are Riesz bases.*

3.2.4 Preconditioning

Here we briefly summarize how the Fast Wavelet Transformation $\mathbf{T}_J$ defined in (3.2.23) can be used for preconditioning together with a diagonal scaling induced by the norm equivalence (3.2.61) in the situation of uniform refinements, i.e., when $S(\Phi_J) = S(\Psi^J)$ [DK1].

To this end, we place ourselves again into the situation of Section 2 for the case $M = 1$. (The case $M > 1$ is treated accordingly by applying the norm equivalence to each of the different function norms). Note that the norm equivalence (3.2.61) implies that every $v \in H^s = H_{1,0}$ can be expanded uniquely in terms of the $\psi_{j,k}$ and its expansion coefficients catenated in $\mathbf{v}$ satisfy

$$\|v\|_{H_{1,0}} \sim \|\mathbf{D}\mathbf{v}\|_{\ell_2}$$

where $\mathbf{D}$ is a diagonal matrix with entries $\mathbf{D}_{(j,k),(j',k')} = 2^{sj}\delta_{j,j'}\delta_{k,k'}$.

In a stable Galerkin scheme of the form (2.36) for $S(\Psi^J) = S(\Psi_\Lambda)$, we have therefore already identified the diagonal (scaling) matrix $\mathbf{D}_J$ consisting of the finite portion of the matrix $\mathbf{D}$ for which $j_0 - 1 \leq j \leq J - 1$. Furthermore, one can check that the representation of $\mathcal{L}$ with respect to the wavelet basis Ψ^J can be expressed in terms of the Fast Wavelet Transform $\mathbf{T}_J$, that is,

$$\langle \Psi^J, \mathcal{L}\Psi^J \rangle = \mathbf{T}_J^T \langle \Phi_J, \mathcal{L}\Phi_J \rangle \mathbf{T}_J, \tag{3.2.74}$$

where Φ_J is the single–scale basis for $S(\Psi^J)$. Thus, one can first set up the operator equation as in Finite Element settings in terms of the single–scale basis Φ_J. Applying the Fast Wavelet Transform $\mathbf{T}_J$ together with $\mathbf{D}_J$ yields that the operator

$$\mathbf{L}_J := \mathbf{D}_J^{-1} \mathbf{T}_J^T \langle \Phi_J, \mathcal{L}\Phi_J \rangle \mathbf{T}_J \mathbf{D}_J^{-1} \tag{3.2.75}$$

has uniformly bounded condition numbers (2.37) independent of J.

Numerical tests confirm that the absolute constants in the estimates (2.37) can further be improved by taking instead of $\mathbf{D}_J^{-1}$ the inverse of the diagonal of $\langle \Psi^J, \mathcal{L}\Psi^J \rangle$ for the scaling in (3.2.75) [CM2, KPV], see e.g. Tables 3.1 and 3.2 taken from [KPV]. These tables display the condition numbers for discretizations of an operator on a one–dimensional domain with periodic boundary conditions where periodized biorthogonal B–spline–wavelets based on the construction from [CDF] are used. The coarsest level j_0 is chosen as $j_0 = 0$. Furthermore, $d, \tilde{d}$ denote the order of the primal and dual multiresolution analyses $\mathcal{S}$ and $\tilde{\mathcal{S}}$. The growth of the condition numbers without preconditioning depending on the finest level J can be seen in the second column ('x' indicates that the condition number has not been computed). In the third column, the condition numbers are computed using the scaling by $\mathbf{D}_J^{-1}$ as in (3.2.75). The last column shows slightly better constants; here the inverse of the diagonal of $\langle \Psi^J, \mathcal{L}\Psi^J \rangle$ is used for the scaling in (3.2.75). The two tables also reveal that raising the order of the multiresolution analyses does not change the constants substantially.

For wavelets on domains discussed below, it is known that the boundary adaptations aggravate the absolute values of the condition numbers. These numbers can be greatly reduced by sophisticated biorthogonalizations of the boundary adapted functions [Ba].

3.2.5 An Example of Biorthogonal Wavelets on $\mathbb{R}$

As an example for the framework presented in Section 3.2.1, consider the situation from [CDF] for $L_2 = L_2(\mathbb{R})$. There the $\phi_{j,k}$ are generated through the dilates and translates of a single function $\phi \in L_2$,

$$\phi_{j,k} = 2^{j/2}\phi(2^j \cdot -k). \tag{3.2.76}$$

Level J	no prec.	$\mathbf{D}_J^{-1}$	$(\text{diag}(\langle\Psi^J, \mathcal{L}\Psi^J\rangle))^{-1}$
3	2.73e+02	5.17e+00	1.62e+00
4	1.11e+03	5.59e+00	1.95e+00
5	4.43e+03	6.13e+00	2.27e+00
6	1.77e+04	6.50e+00	2.51e+00
7	7.08e+04	6.83e+00	2.72e+00
8	2.83e+05	7.09e+00	2.88e+00
9	x	7.32e+00	3.01e+00

Table 3.1: Condition numbers for $\mathcal{L}u = -\Delta u + u$ using periodized biorthogonal spline–wavelets, $j_0 = 0$, $d = \tilde{d} = 2$.

Level J	no prec.	$\mathbf{D}_J^{-1}$	$(\text{diag}(\langle\Psi^J, \mathcal{L}\Psi^J\rangle))^{-1}$
3	3.43e+02	5.51e+00	4.77e+00
4	1.37e+03	5.51e+00	4.88e+00
5	5.46e+03	5.51e+00	4.91e+00
6	2.18e+04	5.51e+00	4.92e+00
7	8.74e+04	5.51e+00	4.92e+00
8	3.49e+05	5.51e+00	4.92e+00
9	x	5.51e+00	4.92e+00

Table 3.2: Condition numbers for $\mathcal{L}u = -\Delta u + u$ using periodized biorthogonal spline–wavelets, $j_0 = 0$, $d = \tilde{d} = 3$.

This corresponds to the idea of a *uniform* virtual underlying grid, explaining the terminology *uniform refinements.* B–Splines on uniform grids are known to satisfy refinement relations (3.2.7) in addition to being compactly supported and having L_2–stable integer translates. For computations, they have the additional advantage that they can be expressed as piecewise polynomials. In the context of variational formulations for second order boundary value problems, a well–used example are the nodal finite elements $\phi_{j,k}$ generated by the cardinal B–Spline of order two, i.e., the piecewise linear continuous function commonly called the 'hat function'. For cardinal B–Splines as generators, a whole class of dual generators $\tilde{\phi}_{j,k}$ (of arbitrary smoothness at the expense of larger supports) can be constructed which are also generated by one single function $\tilde{\phi}$ through translates and dilates. By Fourier techniques, one can construct from $\phi, \tilde{\phi}$ then a pair of biorthogonal wavelets $\psi, \tilde{\psi}$ whose dilates and translates built as in (3.2.76) constitute Riesz bases for $L_2(\mathbb{R})$.

By taking tensor products of these functions, of course, one can generate biorthogonal wavelet bases for $L_2(\mathbb{R}^n)$.

3.3 Wavelets on an Interval

Since operator equations typically live on bounded domains or manifolds, one needs to construct wavelet bases there. Some constructions that exist by now have as a core ingredient tensor products of one-dimensional wavelets on an *interval* derived from the biorthogonal wavelets from [CDF] on $\mathbb{R}$. On finite intervals in $\mathbb{R}$, the corresponding constructions are usually based on keeping the elements of $\Phi_j, \tilde{\Phi}_j$ supported *in* the interval while modifying those translates overlapping the end points of the interval so as to preserve a desired degree of polynomial exactness.

Depending on the type of application, one might not explicitly need the polynomial exactness for the dual function spaces on all of the interval. This is part of the ideas pursued in [AHJP] where only the primal functions are adapted to an interval such that they locally reproduce polynomials up to a certain degree. However, for integral equations one needs in addition a certain number of vanishing moments for the primal wavelets [DPS1]. This requires that the dual multiresolution has a certain order of approximation. A general detailed construction satisfying all these requirements has been proposed in [DKU2], see also [LT, Mal]. Here just the main ideas for constructing a biorthogonal pair $\Phi_j, \tilde{\Phi}_j$ and corresponding wavelets satisfying the above requirements are sketched, where we apply the techniques from Section 3.2.1.

One starts out with those functions from two collections of biorthogonal generators $\Phi_j^{\mathbb{R}}, \tilde{\Phi}_j^{\mathbb{R}}$ for some fixed $j \geq j_0$ living on the whole real line whose support has nonempty intersection with the interval $(0,1)$. In order to treat the boundary effects separately, it is assumed that the coarsest resolution level j_0 is large enough so that, in view of (3.2.5), functions overlapping one end of the interval vanish at the other. One then leaves as many functions from the collection $\Phi_j^{\mathbb{R}}, \tilde{\Phi}_j^{\mathbb{R}}$ living in the interior of the interval untouched and modifies only those near the interval ends. Keeping just the restrictions to the interval of those translates overlapping the end points would destroy stability (and also the cardinality of the primal and dual basis functions living on $(0,1)$ since their supports do not have the same size). Therefore, modifications over the end points are necessary; also, just discarding them from the collections (3.2.3), (3.2.34) would produce an error near the ends of the interval. The basic idea is essentially the same for all constructions of orthogonal and biorthogonal wavelets on $\mathbb{R}$ adapted to an interval. Namely, one takes *fixed* linear combinations of all functions in $\Phi_j^{\mathbb{R}}, \tilde{\Phi}_j^{\mathbb{R}}$ living near the ends of the interval in such a way that monomials up to the exactness order are reproduced there and such that the generator bases have the same cardinality. Because of the boundary modifications, the collections of generators are there no longer biorthogonal. However, one can show in the case of cardinal B–Splines as primal generators (which is the most widely used class for numerical analysis) that biorthogonalization is indeed possible [DKU2]. This yields collections denoted by $\Phi_j^{(0,1)}, \tilde{\Phi}_j^{(0,1)}$ which then satisfy (3.2.35) on $(0,1)$ and all assumptions required in Proposition 3.3.

For the construction of corresponding wavelets, first an *initial* stable completion $\check{\mathbf{M}}_{j,1}$ is computed by applying Gaussian eliminations to factor $\mathbf{M}_{j,0}$ and then to find a uniformly stable inverse of $\check{\mathbf{M}}_j$. Here we exploit that for cardinal B–Splines as generators the refinement matrices $\mathbf{M}_{j,0}$ are totally positive. Thus, they can be stably decomposed by Gaussian elimination without pivoting, a fact that has been used already in [DM1].

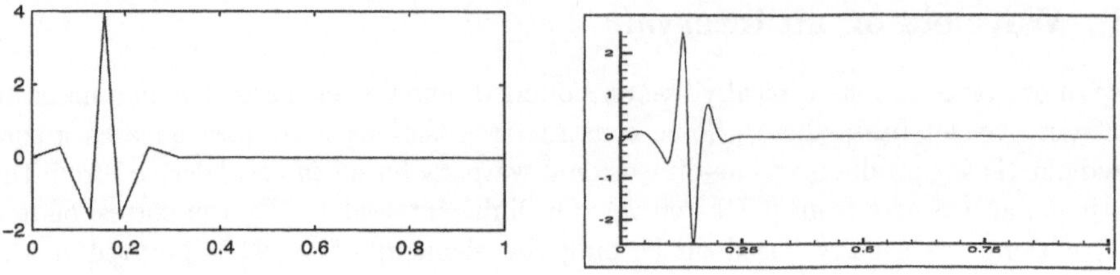

Figure 3.1: Examples of primal wavelets for $d = 2, 3$ adapted to the interval.

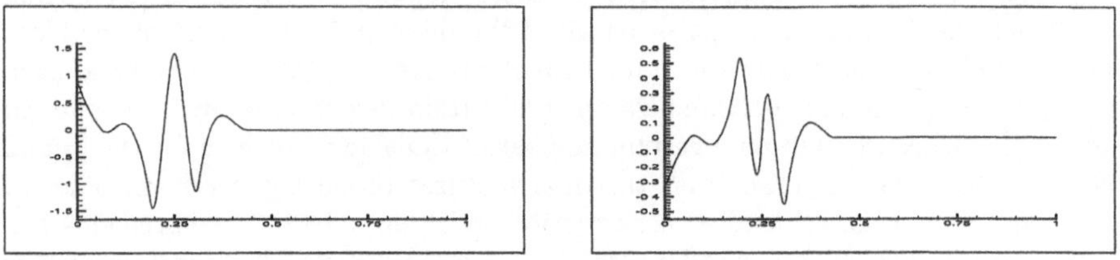

Figure 3.2: Primal multiwavelets adapted to the interval.

Applying then Proposition 3.3 gives the corresponding biorthogonal wavelets $\Psi_j^{(0,1)}$, $\tilde{\Psi}_j^{(0,1)}$ on $(0, 1)$ which satisfy the requirements in Corollary 3.7. It turns out that these wavelets coincide in the interior of the interval again with those on all of $I\!R$ from [CDF]. Two examples of the primal wavelets are displayed in Figure 3.1 for $d = 2$ (generated by piecewise linear continuous functions) and $d = 3$ (generated by cubic B–splines).

After constructing these basic versions, one can then perform local transformations near the ends of the interval in order to improve the condition or L_2 stability constants, see [DKU3] for corresponding results and numerical examples.

It is also possible to construct biorthogonal generators and wavelets with homogeneous (Dirichlet) boundary conditions. This can be done as follows. Since the $\Phi_j^{(0,1)}$ are locally near the boundary monomials which all vanish at $0, 1$ except for one, removing the one from $\Phi_j^{(0,1)}$ which corresponds to the constant function produces a collection of generators with homogeneous boundary conditions at $0, 1$. In order for the moment conditions (3.2.64) still to hold for the Ψ_j, the dual generators have to have *complementary* boundary conditions. A corresponding construction has been carried out in [DS2]. Homogeneous boundary conditions of higher order can be generated accordingly.

For fourth order problems, one could also employ the biorthogonal multiwavelets on the interval based on the C^1 Hermite cubics as generators constructed in [DHJK], see Figure 3.3 for two examples of primal multiwavelets. These functions have the additional advantage of *interpolating* function values and first derivatives at the expense of requiring two sets of primal and dual generators each.

By taking tensor products of the wavelets on $(0, 1)$, one can generate biorthogonal wavelets on $(0, 1)^n$ or any other domain which is the image of a regular parametric mapping of the unit cube with the desired type of boundary conditions on each of its faces. This yields in particular biorthogonal wavelet bases $\Psi_j^{\square}, \tilde{\Psi}_j^{\square}$ for $H^s(\square)$.

Wavelets for Sobolev spaces on boundary manifolds $H^s(\Gamma)$, $\Gamma \subseteq \partial\Omega$, are discussed next.

3.4 Wavelets on Manifolds

We briefly recall some results on the construction of wavelets on manifolds summarized in [D4].

3.4.1 Domain Decomposition Approaches

These (nonoverlapping) domain decomposition approaches apply to manifolds Γ which can be represented as the union of (essentially) disjoint smooth parametric images of the open cube

$$\square^{n-1} := (0,1)^{n-1} \tag{3.4.1}$$

which denotes the standard parameter domain. That is, with smooth parametrizations $\kappa^i : \mathbb{R}^{n-1} \to \mathbb{R}^n$, $i = 1, \ldots, K$, let Γ have the representation

$$\overline{\Gamma} = \bigcup_{i=1}^{K} \overline{\Gamma^i}, \quad \Gamma^i = \kappa^i(\square^{n-1}), \quad i = 1, \ldots, K, \tag{3.4.2}$$

where $\Gamma^i \cap \Gamma^k = \emptyset$ for $i \neq k$. For instance, an L–shaped manifold in the plane can be decomposed as shown in Figure 3.4.1. Let $\Psi^{\square^{n-1}}$ be the wavelets constructed on the

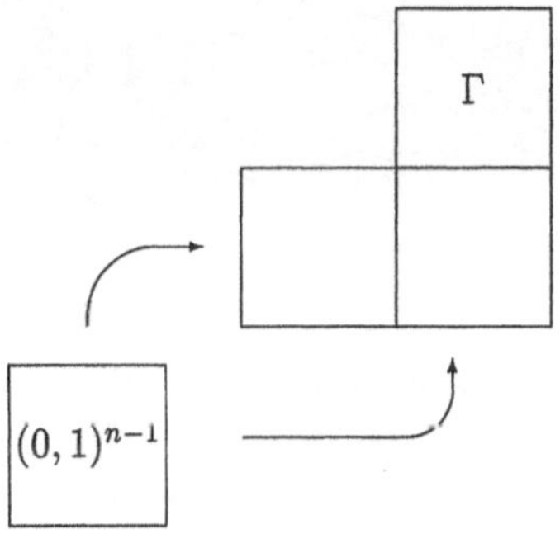

Figure 3.3: Decomposition of an L–shaped manifold.

simple domain $\square^{n-1}$ by tensor products from Section 3.3. Wavelets on Γ^i defined by

$$\Psi^{\Gamma^i} := \Psi^{\square^{n-1}} \circ (\kappa^i)^{-1} := \{\psi^{\square^{n-1}} \circ (\kappa^i)^{-1} : \psi^{\square^{n-1}} \in \Psi^{\square^{n-1}}\} \tag{3.4.3}$$

and $\tilde{\Psi}^{\Gamma^i} := \tilde{\Psi}^{\square^{n-1}} \circ (\kappa^i)^{-1}$ are then biorthogonal with respect to the modified inner product

$$(v, w)_{\mathrm{mod}(L_2(\Gamma^i))} := \int_{\square^{n-1}} v(\kappa^i(\mathbf{x}))\, w(\kappa^i(\mathbf{x}))\, d\mathbf{x}. \tag{3.4.4}$$

The constructions in [CTU1, CTU2, CM1, DS1, JL, Ma2] yield *globally continuous* biorthogonal wavelets $\Psi^\Gamma, \tilde{\Psi}^\Gamma$ where biorthogonality holds relative to the inner product

$$(v,w)_{\mathrm{mod}(L_2(\Gamma))} := \sum_{i=1}^{K} (v|_{\Gamma^i}, w|_{\Gamma^i})_{\mathrm{mod}(L_2(\Gamma^i))} \tag{3.4.5}$$

with $(\cdot,\cdot)_{\mathrm{mod}(L_2(\Gamma^i))}$ defined in (3.4.4), where $(\cdot)|_{\Gamma^i}$ denotes restriction to Γ^i.

The construction of these wavelets is based on the following steps. First globally continuous (primal and dual) generators $\Phi^\Gamma, \tilde{\Phi}^\Gamma$ are formed by 'gluing' them together across patch boundaries. This is relatively easy since the functions in $\Phi^{\square^{n-1}}, \tilde{\Phi}^{\square^{n-1}}$ are built from univariate functions which all vanish at the boundary except for one at each end. Different strategies have then be used in [CTU1] and [DS1] to construct the wavelets $\Psi^\Gamma, \tilde{\Psi}^\Gamma$. What they all have in common is the fact that the norm equivalences (3.2.61) for $H^s = H^s(\Gamma)$ can be shown to hold for the range $-1/2 < s < 3/2$. This is due to the fact that duality arguments apply only for this range because of the nature of the modified inner product (3.4.5) to which biorthogonality refers.

An example of a primal wavelet generated by continuous piecewise linear functions which is constructed at the corner of a box is displayed in Figure 3.4 [DHaS].

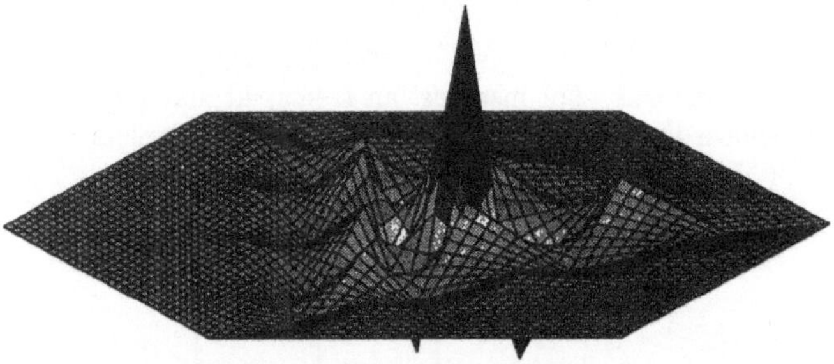

Figure 3.4: Piecewise linear continuous wavelet at the corner of a box.

Numerical results for solutions of second order elliptic problems on two-dimensional L-shaped domains based on the construction [DS1] can be found in [Vo]; corresponding results for the construction [CTU1] also for three-dimensional domains are provided in [CTU2, CTU3].

In view of Section 4, a wavelet basis for $(H^{1/2}(\Gamma))'$ (which is $H^{-1/2}(\partial\Omega)$ for $\Gamma = \partial\Omega$) would be needed for which the above constructions would so far not be sufficient since $s = -1/2$ is excluded. However, one can still use these constructions as follows. Consider the modified inner product (3.4.5) and wavelets Ψ^{Γ^i}, $\tilde{\Psi}^{\Gamma^i}$ defined in (3.4.3) which are biorthogonal with respect to (3.4.4). Note that there exists a positive piecewise smooth function ϱ which is smooth on each patch such that by substitution the identity

$$(v,w)_{\mathrm{mod}(L_2(\Gamma))} = (\varrho v, w)_{L_2(\Gamma)}$$

holds, i.e., Ψ^{Γ^i} and $\varrho\tilde{\Psi}^{\Gamma^i}$ are biorthogonal with respect to the standard inner product $(\cdot,\cdot)_{L_2(\Gamma)}$. If then Ψ^{Γ^i} satisfies the norm equivalence (3.2.61) for $s = 1/2$, it follows by duality that for $\varrho\tilde{\Psi}^{\Gamma^i}$ the norm equivalence (3.2.61) holds for $s = -1/2$. Since it will not be needed for the applications below that $\varrho\tilde{\Psi}^{\Gamma^i}$ is globally continuous, this construction would be sufficient for the situations considered in the subsequent chapters.

3.4.2 The Cartesian Product Approach

An alternative concept to construct wavelet bases on Γ which overcomes the above limitations on the ranges for which the norm equivalences hold has been developed in [DS3]. It is still based on the decomposition (3.4.2) of Γ. However, it differs essentially from the above approach in that the construction of bases is closely related to the characterization of functions spaces defined on Γ as *Cartesian products* of certain local spaces based on the partition of Γ and not as usual on an atlas and charts. This characterization has been given first in [CF] for a wide range of Besov and Sobolev spaces in order to construct unconditional bases for function spaces on compact C^∞–manifolds. Also an explicit construction has been given there based on spline bases arising from Gram–Schmidt orthogonalizations. Because of their lack of locality, the objective in [DS3] was to adapt the concepts from [CF] to generate compactly supported biorthogonal systems for applications in numerical analysis.

We briefly sketch the ideas of the construction from [DS3, D4], restricting ourselves to Sobolev spaces. The main point is that $H^s(\Gamma)$ is equivalent to the *product* of certain closed subspaces of the Sobolev spaces on each patch $H^s(\Gamma^i)$.

To describe this, first a numbering of the patches Γ^i is fixed and a direction is assigned to each interface $\overline{\Gamma}^i \cap \overline{\Gamma}^j$. According to this direction, the *inflow* and *outflow boundary* contains those faces of Γ^i whose directions point into, respectively, out of Γ^i. Now let $\Gamma^{i,\uparrow}$ denote a slightly larger patch in Γ which contains Γ^i such that the inflow boundary is also part of the boundary of $\Gamma^{i,\uparrow}$. The patch $\Gamma^{i,\downarrow}$ is defined correspondingly for reversed directions. Consider then the following extension from Γ^i to $\Gamma^{i,\uparrow}$. Define $H^s(\Gamma^i)^\uparrow := \{v \in H^s(\Gamma^i) : \chi_{\Gamma^i} v \in H^s(\Gamma^{i,\uparrow})\}$ with corresponding norm $\|v\|_{H^s(\Gamma^i)^\uparrow} := \|\chi_{\Gamma^i} v\|_{H^s(\Gamma^{i,\uparrow})}$, where χ_{Γ^i} is the characteristic function on Γ^i. Thus, the space $H^s(\Gamma^i)^\uparrow$ contains those functions whose trivial extension by zero across the outflow boundary has the same Sobolev regularity on the larger domain $\Gamma^{i,\uparrow}$. Note that for $s \neq n + 1/2$, n an integer, these spaces agree with $H^s_{0,\partial\Gamma^{i,\uparrow}}(\Gamma^i) := \{v \in H^s(\Gamma^i) : v|_{\partial\Gamma^{i,\uparrow}} = 0\}$ (see [Gr], Corollary 1.4.4.5). In this sense, the functions in $H^s(\Gamma^i)^\uparrow$ satisfy certain *homogeneous boundary conditions* at the outflow boundary. Analogously, $H^s(\Gamma^i)^\downarrow$ is defined. On the basis of appropriate bounded extension operators $E_i : H^s(\Gamma^i) \to H^s(\Gamma^{i,\uparrow})$ whose adjoints E_i' are also continuous, one can construct projectors $P_i : H^s(\Gamma) \to H^s(\Gamma)$ such that

$$(P_i(H^s(\Gamma)))|_{\Gamma^i} = H^s(\Gamma^i)^\downarrow, \qquad (P_i'(H^s(\Gamma)))|_{\Gamma^i} = H^s(\Gamma^i)^\uparrow \tag{3.4.6}$$

holds. Then one can show that the topological isomorphisms

$$\begin{aligned}
H^s(\Gamma) &\simeq \prod_{i=1}^K H^s(\Gamma^i)^{\downarrow}, & \|v\|_{H^s(\Gamma)} &\sim \left(\sum_{i=1}^K \|(P_i v)_{\Gamma^i}\|^2_{H^s(\Gamma^i)^{\downarrow}}\right)^{1/2}, \\
(H^s(\Gamma))' &\simeq \prod_{i=1}^K H^{-s}(\Gamma^i)^{\uparrow}, & \|v\|_{(H^s(\Gamma))'} &\sim \left(\sum_{i=1}^K \|(P_i' v)_{\Gamma^i}\|^2_{H^{-s}(\Gamma^i)^{\uparrow}}\right)^{1/2},
\end{aligned} \tag{3.4.7}$$

are valid, where $H^{-s}(\Gamma^i)^{\uparrow}$ is the dual of $H^s(\Gamma^i)^{\downarrow}$. Based on (3.4.7), one can now construct local wavelet bases

$$\Psi^{\Gamma^i} := \Psi^{\square^{i,\downarrow}} \circ (\kappa^i)^{-1}, \qquad \tilde{\Psi}^{\Gamma^i} := \tilde{\Psi}^{\square^{i,\uparrow}} \circ (\kappa^i)^{-1}, \tag{3.4.8}$$

where $\square^{i,\downarrow}$ indicates that the wavelets in $\Psi^{\square^{i,\downarrow}}$ satisfy those boundary conditions induced by the inflow and outflow boundaries on Γ^i through the parametric pull back $(\kappa^i)^{-1}$. Specifically, the functions from the dual collection $\tilde{\Psi}^{\Gamma^i}$ satisfy corresponding *complementary* boundary conditions as indicated by the reverse arrow $\uparrow$. Thus, the construction (3.4.8) is reduced to the construction of biorthogonal wavelet bases on the unit cube with complementary boundary conditions, and the lifted bases $\Psi^{\Gamma^i}, \tilde{\Psi}^{\Gamma^i}$ satisfy then norm equivalences

$$\|v\|_{H^s(\Gamma^i)^{\downarrow}} \sim \|\{2^{sj}(v, \tilde{\psi}^{\Gamma^i}_{j,k})_{L_2(\Gamma^i)}\}\|_{\ell_2}, \tag{3.4.9}$$

see [DS2]. The global wavelets $\Psi^{\Gamma}, \tilde{\Psi}^{\Gamma}$ are now formed as follows. By assumption, the function $g_i := |\partial\kappa^i((\kappa^i)^{-1})|$ is smooth on Γ^i. The indices λ for $\Psi^{\Gamma}, \tilde{\Psi}^{\Gamma}$ have the form $\lambda = (\lambda_i, i)$ where λ_i (corresponding to (j,k) in (3.4.9)) is an index for the basis Ψ^{Γ^i}. Setting for any such λ

$$\psi^{\Gamma}_{\lambda} := P_i(\chi_{\Gamma^i}\psi^{\Gamma^i}_{\lambda_i,i}), \qquad \tilde{\psi}^{\Gamma}_{\lambda} := P_i'(\chi_{\Gamma^i} g_i^{-1}\tilde{\psi}^{\Gamma^i}_{\lambda_i,i}), \tag{3.4.10}$$

the factor g_i^{-1} in the definition of the dual basis functions can be seen to ensure biorthogonality with respect to the canonical inner product $(\cdot,\cdot)_{L_2(\Gamma)}$, i.e.,

$$(\Psi^{\Gamma}, \tilde{\Psi}^{\Gamma})_{L_2(\Gamma)} = \mathbf{I}. \tag{3.4.11}$$

Furthermore, combining the local norm equivalences (3.4.9) with (3.4.7) yields

$$\|v\|_{H^s(\Gamma)} \sim \|\mathbf{D}\,\langle v, \tilde{\Psi}^{\Gamma}\rangle^T\|_{\ell_2} \tag{3.4.12}$$

where $\mathbf{D}_{(j,k),(j',k')} = 2^{sj}\delta_{(j,k),(j',k')}$. The range for $s \in (-\tilde{\sigma}, \sigma)$ for which (3.4.12) holds depends on the regularity of the wavelets $\Psi^{\square^{i,\downarrow}}, \tilde{\Psi}^{\square^{i,\uparrow}}$ as well as on the regularity of Γ and the type of extension.

Note that in contrast to the constructions in Section 3.4.1, biorthogonality holds here with respect to the *canonical* inner product, implying that duality arguments apply in the full range of validity in the norm equivalence (3.4.12). Moreover, only the regularity of Γ restricts in principle the validity of (3.4.12). Also it should be mentioned that one does not need a global parametrization for Γ. It is interesting to note that the construction based on Cartesian products also induces automatically a domain decomposition scheme for global *operators* such as integral operators, see [DS3] for details.

These concepts provide in principle a convenient way to construct piecewise linear continuous wavelets $\Psi^{\square^{i,\downarrow}}$, $\tilde{\Psi}^{\square^{i,\uparrow}}$ satisfying globally the norm equivalence (3.2.61) for the range $|s| \leq 1$ or smoother variants permitted by the regularity of the manifold. For the subsequent chapters, one actually needs less, see the remarks at the end of Section 3.4.1.

3.5 Multiscale Decomposition of Function Spaces — Non–Uniform Refinements

In the concept of multiresolution analyses of L_2 in Section 3.2 the generators and wavelets have been indexed by a pair of indices (j, k) where j stands for the resolution level and k contains information on the spatial location and the type of wavelet. An example where this notation has been useful are the Jackson estimates (3.2.55) which correspond to estimates in spaces on uniform grids. Nevertheless, one will also want to work with *arbitrary* subsets in which case a change of labelling is more convenient. We will use then indices λ to label functions and $|\lambda|$ to refer to the refinement level.

Here we briefly list a few facts from Section 3.2 again in this notation which is more convenient for later reference.

Choosing a finite–dimensional set $\Lambda \subset I\!I$ we set

$$S_\Lambda = S(\Psi_\Lambda) = S(\{\psi_\lambda : \lambda \in \Lambda\}), \quad \tilde{S}_\Lambda = S(\tilde{\Psi}_\Lambda) = S(\{\tilde{\psi}_\lambda : \lambda \in \Lambda\}),$$

and

$$P_\Lambda v = \langle v, \tilde{\Psi}_\Lambda \rangle \Psi_\Lambda, \quad P'_\Lambda v = \langle v, \Psi_\Lambda \rangle \tilde{\Psi}_\Lambda,$$

denote the corresponding projectors onto $S_\Lambda, \tilde{S}_\Lambda$. In particular,

$$S_\Lambda \subset S(\Phi_J) \quad \text{for } J = \max\{|\lambda| + 1 : \ \lambda \in \Lambda\},$$

and for $\hat{\Lambda} \supset \Lambda$ one has $S_\Lambda \subset S_{\hat{\Lambda}}$. Since the set of wavelets Ψ_Λ is a subset of Ψ defined in (3.2.29), its component functions still satisfy a refinement relation (3.2.12).

Of course, one still has the biorthogonality relation (3.2.32),

$$\langle \Psi_\Lambda, \tilde{\Psi}_\Lambda \rangle = \mathbf{I},$$

and any function $v \in S_\Lambda$ has the expansion

$$v = \langle v, \tilde{\Psi}_\Lambda \rangle \Psi_\Lambda = \sum_{\lambda \in \Lambda} \langle v, \tilde{\psi}_\lambda \rangle \psi_\lambda.$$

In the case of non–uniform refinements, one cannot expect the Jackson estimates (3.2.55) to hold.

In this notation the discrete norms $|\!|\!| \cdot |\!|\!|_s$ from (3.2.66) can be rewritten as

$$|\!|\!| v |\!|\!|_s = \left(\sum_{\lambda \in I\!I} 2^{2s|\lambda|} |\langle v, \tilde{\psi}_\lambda \rangle|^2 \right)^{1/2}.$$

4 Elliptic Boundary Value Problems

The aim of this chapter is to lay the foundation for the control problems discussed in Chapter 6 where the boundary control plays an essential role. Here first elliptic boundary value problems are treated where particular emphasis is placed on the appropriate flexible treatment of essential non–homogeneous boundary conditions. An approach meeting this goal is to formulate the elliptic boundary value problem as a *saddle point problem* in Section 4.2.1.

In view of this problem, we recall beforehand a few facts about saddle point problems in an abstract framework from e.g. [BF].

4.1 General Saddle Point Problems

4.1.1 The Continuous Abstract Saddle Point Problem

Let $H_{1,0} = Y$ and $H_{2,0} = Q$ be Hilbert spaces with their topological duals Y', Q' and dual forms $\langle\cdot,\cdot\rangle_{Y\times Y'}$, $\langle\cdot,\cdot\rangle_{Q\times Q'}$, respectively. We denote the norms on Y and Q and the induced inner products by

$$(\cdot,\cdot)_Y = \|\cdot\|_Y^2 \quad \text{and} \quad (\cdot,\cdot)_Q = \|\cdot\|_Q^2 \,. \tag{4.1.1}$$

Let $a(\cdot,\cdot)$ be a continuous bilinear form defined on $Y \times Y$,

$$a(v,w) \leq \tilde{\alpha}_2 \|v\|_Y \|w\|_Y, \qquad v,w \in Y, \qquad \tilde{\alpha}_2 > 0, \tag{4.1.2}$$

which defines a linear continuous operator $A : Y \to Y'$ by

$$\langle Av, w\rangle_{Y'\times Y} := a(v,w), \qquad v,w \in Y. \tag{4.1.3}$$

Let there be a second bilinear form $b(\cdot,\cdot)$ on $Y \times Q$ which is also continuous,

$$b(v,q) \leq \tilde{\beta}_2 \|v\|_Y \|q\|_Q, \qquad v \in Y,\ q \in Q, \qquad \tilde{\beta}_2 > 0. \tag{4.1.4}$$

A linear operator $B : Y \to Q'$ and its adjoint $B' : Q \to Y'$ can then be defined by

$$\langle Bv, q\rangle_{Q'\times Q} = \langle v, B'q\rangle_{Y\times Y'} := b(v,q), \qquad v \in Y,\ q \in Q. \tag{4.1.5}$$

Furthermore, we identify

$$\mathcal{H} = Y \times Q, \qquad \left\| \begin{pmatrix} v \\ q \end{pmatrix} \right\|_{\mathcal{H}}^2 = \|v\|_Y^2 + \|q\|_Q^2, \tag{4.1.6}$$

according to (2.4). Define also for $q \in Q$

$$\|q\|_{Q/\ker B'} := \inf_{q_0 \in \ker B'} \|q + q_0\|_Q. \tag{4.1.7}$$

In particular, if B is surjective, one has $\ker B' = \{0\}$ implying $\|q\|_{Q/\ker B'} = \|q\|_Q$.

We can now formulate the general problem: Given $(f,u) \in \mathcal{H}'$, find $(y,p) \in \mathcal{H}$ such that

$$\begin{aligned} a(y,v) + b(v,p) &= \langle f, v\rangle_{Y'\times Y} \qquad \text{for all } v \in Y, \\ b(y,q) &= \langle u, q\rangle_{Q'\times Q} \qquad \text{for all } q \in Q. \end{aligned} \tag{4.1.8}$$

In operator form, this can be rephrased as follows: Given $(f,u) \in \mathcal{H}'$, find $(y,p) \in \mathcal{H}$ such that

$$\begin{pmatrix} A & B' \\ B & 0 \end{pmatrix} \begin{pmatrix} y \\ p \end{pmatrix} = \begin{pmatrix} f \\ u \end{pmatrix} \tag{4.1.9}$$

holds. Concerning the existence and uniqueness of solutions of general saddle point problems, we recall the following result from e.g. [Br, BF, GR].

Theorem 4.1 *Let the linear operator A be invertible on* $\ker B \subseteq Y$*, i.e., for some constant $\tilde{\alpha}_1 > 0$*

$$\begin{aligned} \inf_{v\in\ker B} \sup_{w\in\ker B} \frac{\langle Av, w\rangle_{Y'\times Y}}{\|v\|_Y\|w\|_Y} &\geq \tilde{\alpha}_1, \\ \inf_{v\in\ker B} \sup_{w\in\ker B} \frac{\langle A'v, w\rangle_{Y'\times Y}}{\|v\|_Y\|w\|_Y} &\geq \tilde{\alpha}_1, \end{aligned} \tag{4.1.10}$$

and let the range of B be closed in Q', i.e., for some constant $\tilde{\beta}_1 > 0$ the inf-sup condition

$$\sup_{v\in Y} \frac{\langle Bv, q\rangle_{Q'\times Q}}{\|v\|_Y} \geq \tilde{\beta}_1 \|q\|_{Q/\ker B'}, \quad q \in Q, \tag{4.1.11}$$

holds. Then there exists a unique solution $(y,p) \in \mathcal{H}$ to problem (4.1.9) for given $(f,u) \in \mathcal{H}'$, i.e.,

$$\mathcal{L} := \begin{pmatrix} A & B' \\ B & 0 \end{pmatrix} : \mathcal{H} \to \mathcal{H}' \tag{4.1.12}$$

is an isomorphism, and one has the equivalence

$$c_{\mathcal{L}} \left\| \begin{pmatrix} v \\ q \end{pmatrix} \right\|_{\mathcal{H}} \leq \left\| \mathcal{L} \begin{pmatrix} v \\ q \end{pmatrix} \right\|_{\mathcal{H}'} \leq C_{\mathcal{L}} \left\| \begin{pmatrix} v \\ q \end{pmatrix} \right\|_{\mathcal{H}} \tag{4.1.13}$$

for any $(v,q) \in \mathcal{H}$, where the constants $c_{\mathcal{L}}, C_{\mathcal{L}}$ are given as

$$\begin{aligned} c_{\mathcal{L}} &:= \left(\tfrac{1}{\tilde{\alpha}_1\tilde{\beta}_1}\left(1 + \tfrac{\tilde{\alpha}_2}{\tilde{\alpha}_1}\right) + \max\left\{ \tfrac{2}{\tilde{\alpha}_1^2}, \tfrac{1}{\tilde{\beta}_1^2}\left(1+\tfrac{\tilde{\alpha}_2}{\tilde{\alpha}_1}\right)^2 + \left(\tfrac{\tilde{\alpha}_2}{\tilde{\beta}_1^2}\left(1+\tfrac{\tilde{\alpha}_2}{\tilde{\alpha}_1}\right)\right)^2 \right\} \right)^{-1/2} \\ C_{\mathcal{L}} &:= \sqrt{2(\tilde{\alpha}_2^2 + \tilde{\beta}_2^2)}. \end{aligned} \tag{4.1.14}$$

The derivation of the upper constant $C_{\mathcal{L}}$ follows immediately from (4.1.6) combined with the continuity of A and B (4.1.2) and (4.1.4). As for the estimate from below, it was proved in [Br, BF] that for every $(v,q) \in \mathcal{L}$ the estimates

$$\begin{aligned} \|v\|_Y &\leq \tfrac{1}{\tilde{\alpha}_1}\|Av + B'q\|_{Y'} + \tfrac{1}{\tilde{\beta}_1}\left(1+\tfrac{\tilde{\alpha}_2}{\tilde{\alpha}_1}\right)\|Bv\|_{Q'}, \\ \|q\|_Q &\leq \tfrac{1}{\tilde{\beta}_1}\left(1+\tfrac{\tilde{\alpha}_2}{\tilde{\alpha}_1}\right)\|Av + B'q\|_{Y'} + \tfrac{\tilde{\alpha}_2}{\tilde{\beta}_1^2}\left(1+\tfrac{\tilde{\alpha}_2}{\tilde{\alpha}_1}\right)\|Bv\|_{Q'} \end{aligned} \tag{4.1.15}$$

hold. Squaring and adding both sides yields the representation of the lower constant $c_{\mathcal{L}}$.

Note that the estimates (4.1.15) imply in particular that the unique solution $(y,p) \in \mathcal{H}$ of (4.1.2) satisfies the *a-priori error estimates*

$$\begin{aligned} \|y\|_Y &\le \tfrac{1}{\tilde\alpha_1}\|f\|_{Y'} + \tfrac{1}{\beta_1}\left(1+\tfrac{\tilde\alpha_2}{\tilde\alpha_1}\right)\|u\|_{Q'}, \\ \|p\|_Q &\le \tfrac{1}{\beta_1}\left(1+\tfrac{\tilde\alpha_2}{\tilde\alpha_1}\right)\|f\|_{Y'} + \tfrac{\tilde\alpha_2}{\beta_1^2}\left(1+\tfrac{\tilde\alpha_2}{\tilde\alpha_1}\right)\|u\|_{Q'}. \end{aligned} \tag{4.1.16}$$

For the remainder of the section on general saddle point problems, we require that A satisfies the conditions (4.1.10). We comment on the validity later in special situations. Furthermore, we discuss in this chapter the important case that the operator B is *surjective* so that $\operatorname{range} B = Q'$ and $\ker B' = \{0\}$. Thus, the inf-sup condition (4.1.11) is always satisfied. This case is in accordance with the examples considered later where B is the trace operator.

4.1.2 Discretization and Preconditioning

After identifying $\mathcal{L}$ and $\mathcal{H}$ and ensuring that $\mathcal{L}$ is an isomorphism, we want to apply next Step 2 in Chapter 2 to the case of the general saddle point problem (4.1.9). We have assured that there is a pair of biorthogonal wavelet bases each for both Y and Q so that corresponding versions of the norm equivalences (2.18) and (2.22) hold,

$$\|v\|_Y \sim \|\mathbf{D}_Y \mathbf{v}\|_{\ell_2(I\!I_Y)}, \qquad \|\tilde v\|_{Y'} \sim \|\mathbf{D}_Y^{-1}\tilde{\mathbf{v}}\|_{\ell_2(I\!I_Y)}, \tag{4.1.17}$$

and

$$\|q\|_Q \sim \|\mathbf{D}_Q \mathbf{q}\|_{\ell_2(I\!I_Q)}, \qquad \|\tilde q\|_{Q'} \sim \|\mathbf{D}_Q^{-1}\tilde{\mathbf{q}}\|_{\ell_2(I\!I_Q)}. \tag{4.1.18}$$

All quantities referring to Y or Q will be indexed accordingly by subscript $()_Y$ or $()_Q$, respectively, if distinction is necessary.

Expanding $U = (y,p)^T \in \mathcal{H}$ as in (2.25) in terms of the scaled wavelet bases,

$$U = (y,p)^T = \left(\mathbf{y}^T\mathbf{D}_Y^{-1}\Psi_Y,\, \mathbf{p}^T\mathbf{D}_Q^{-1}\Psi_Q\right)^T =: \mathbf{U}^T\mathbf{D}^{-1}\mathbf{\Psi}, \tag{4.1.19}$$

and also the right hand side $F = (f,u)^T = \langle F, \mathbf{\Psi}\rangle\tilde{\mathbf{\Psi}} \in \mathcal{H}'$ yields after multiplying by $\mathbf{D}^{-1} = \operatorname{diag}(\mathbf{D}_Y^{-1}, \mathbf{D}_Q^{-1})$ as a special case of (2.28) the system of equations

$$\mathbf{L}\begin{pmatrix}\mathbf{y}\\ \mathbf{p}\end{pmatrix} := \begin{pmatrix}\mathbf{A} & \mathbf{B}^T\\ \mathbf{B} & \mathbf{0}\end{pmatrix}\begin{pmatrix}\mathbf{y}\\ \mathbf{p}\end{pmatrix} = \begin{pmatrix}\mathbf{f}\\ \mathbf{u}\end{pmatrix} \tag{4.1.20}$$

where we have used the abbreviations

$$\begin{aligned} \mathbf{A} &:= \mathbf{D}_Y^{-1}\langle\Psi_Y, A\Psi_Y\rangle\mathbf{D}_Y^{-1}, & \mathbf{f} &:= \mathbf{D}_Y^{-1}\langle\Psi_Y, f\rangle, \\ \mathbf{B} &:= \mathbf{D}_Q^{-1}\langle\Psi_Q, B\Psi_Y\rangle\mathbf{D}_Y^{-1}, & \mathbf{u} &:= \mathbf{D}_Q^{-1}\langle\Psi_Q, u\rangle. \end{aligned} \tag{4.1.21}$$

Thus, covered by the general concept from Chapter 2, we have the following result.

Corollary 4.2 *The operator* $\mathbf{L}$ *defined in (4.1.20) is an* ℓ_2*–automorphism, i.e., for every* $(\mathbf{v},\mathbf{q}) \in \ell_2(\mathbb{I}) = \ell_2(\mathbb{I}_Y \times \mathbb{I}_Q)$ *one has*

$$c_{\mathbf{L}} \left\| \begin{pmatrix} \mathbf{v} \\ \mathbf{q} \end{pmatrix} \right\|_{\ell_2} \leq \left\| \mathbf{L} \begin{pmatrix} \mathbf{v} \\ \mathbf{q} \end{pmatrix} \right\|_{\ell_2} \leq C_{\mathbf{L}} \left\| \begin{pmatrix} \mathbf{v} \\ \mathbf{q} \end{pmatrix} \right\|_{\ell_2} \tag{4.1.22}$$

with constants $c_{\mathbf{L}}, C_{\mathbf{L}}$ *only depending on* $c_{\mathcal{L}}, C_{\mathcal{L}}$ *from (4.1.13) and the constants in the norm equivalences (4.1.17) and (4.1.18).*

4.1.3 The Discrete Finite Problem

After Step 2 of the general concept has been completed, we have to investigate under which conditions finite–dimensional systems derived from (4.1.20) are uniformly stable. At this point it is at first more instructive to return from the discretized preconditioned system (4.1.20) to a discretization in standard terms involving again the spaces Y, Q and their norms. In fact, this brings out the precise roles of the involved function spaces for deriving conditions that guarantee stability for general saddle point problems.

To this end, choose a finite subset of indices

$$\Lambda = \Lambda_Y \times \Lambda_Q \subset \mathbb{I}$$

by which the wavelets are labelled according to (2.31). Specifically, we use $\Psi_{Y,\Lambda}$ (shortly for Ψ_{Y,Λ_Y}) and $\Psi_{Q,\Lambda}$ (shortly for Ψ_{Q,Λ_Q}) for the bases of the approximation spaces for Y and Q, denoted by

$$Y_\Lambda := S(\Psi_{Y,\Lambda}), \qquad Q_\Lambda := S(\Psi_{Q,\Lambda}). \tag{4.1.23}$$

Correspondingly, the spaces $\tilde{Y}_\Lambda$, $\tilde{Q}_\Lambda$ are defined as the spans of the dual bases $\tilde{\Psi}_{Y,\Lambda}$ and $\tilde{\Psi}_{Q,\Lambda}$.

The discrete finite problem can then be formulated in operator form as follows: Given (f_Λ, u_Λ), find $(y_\Lambda, p_\Lambda) \in Y_\Lambda \times Q_\Lambda$ such that

$$\begin{pmatrix} A_\Lambda & B_\Lambda^T \\ B_\Lambda & 0 \end{pmatrix} \begin{pmatrix} y_\Lambda \\ p_\Lambda \end{pmatrix} = \begin{pmatrix} f_\Lambda \\ u_\Lambda \end{pmatrix} \tag{4.1.24}$$

holds. In view of the central Theorem 4.1 which, of course, also applies to finite–dimensional spaces, one has to assure that Y_Λ and Q_Λ are chosen such that the bilinear forms $a(\cdot,\cdot)$ and $b(\cdot,\cdot)$ are continuous and satisfy the ellipticity condition (4.1.10) and the inf–sup condition (4.1.11) with respect to the spaces Y_Λ and Q_Λ. The continuity is trivial for any subspaces Y_Λ, Q_Λ of Y, Q so that (4.1.2) and (4.1.4) hold with the same constants $\tilde{\alpha}_2$, $\tilde{\beta}_2$. The *discrete ellipticity condition* reads as follows: The discrete linear operator A_Λ is required to be invertible on $\ker B_\Lambda$ with some constant $\alpha_1 > 0$ independent of Λ,

$$\inf_{v\in\ker B_\Lambda} \sup_{w\in\ker B_\Lambda} \frac{\langle A_\Lambda v, w\rangle_{Y'\times Y}}{\|v\|_Y \|w\|_Y} \geq \alpha_1, \tag{4.1.25}$$

$$\inf_{v\in\ker B_\Lambda} \sup_{w\in\ker B_\Lambda} \frac{\langle A_\Lambda^T v, w\rangle_{Y'\times Y}}{\|v\|_Y \|w\|_Y} \geq \alpha_1.$$

Here we have used

$$\ker B_\Lambda := \{v_\Lambda \in Y_\Lambda : b(v_\Lambda, q_\Lambda) = 0 \text{ for all } q_\Lambda \in Q_\Lambda\}. \tag{4.1.26}$$

Note that (4.1.25) follows trivially from the continuous condition (4.1.10) provided that

$$\ker B_\Lambda \subset \ker B \tag{4.1.27}$$

which is not satisfied in general. In fact, condition (4.1.27) is often called *equilibrium condition* between Y_Λ and Q_Λ, see e.g. [Br].

The analog of (4.1.11) for the pair Y_Λ, Q_Λ is usually referred to as the *Ladyšenskaja–Babuška–Brezzi (LBB) condition* and requires that there exists a constant $\beta_1 > 0$ *independent* of Λ such that

$$\inf_{q\in Q_\Lambda} \sup_{v\in Y_\Lambda} \frac{b(v,q)}{\|v\|_Y \, \|q\|_Q} \geq \beta_1 > 0 \tag{4.1.28}$$

holds.

A general result which follows by biorthogonality (2.24) has been shown in [DHU].

Remark 4.3 *Both the equilibrium (4.1.27) and the LBB condition (4.1.28) are satisfied if Y_Λ and $\tilde{Q}_\Lambda$ are connected by*

$$B(Y_\Lambda) = \tilde{Q}_\Lambda. \tag{4.1.29}$$

As we will see later, condition (4.1.29) is in fact too strong for the elliptic boundary value problems with explicit treatment of boundary conditions discussed below in Section 4.2. Of course, since (4.1.27) is only sufficient for (4.1.25) but not necessary, there are cases where (4.1.25) and (4.1.28) hold without (4.1.29) to be satisfied.

In Section 4.2.4 below we will concentrate on conditions that ensure the discrete inf–sup condition (4.1.28) for the saddle point problems derived in Section 4.2.

4.1.4 Error Estimates and Iterative Solution

Having derived conditions that guarantee stability of the discretizations, it remains to solve the discrete finite system numerically. But first we recall from e.g. [BF] standard error estimates comparing the solution of the continuous problem with the discrete one.

Proposition 4.4 *Let $(y,p) \in Y \times Q$ be the solution of the continuous saddle point problem (4.1.9) and $(y_\Lambda, p_\Lambda) \in Y_\Lambda \times Q_\Lambda$ be the solution of the discrete finite problem (4.1.24). Then the error estimates*

$$\begin{aligned}
\|y - y_\Lambda\|_Y &\leq \left(1 + \frac{\tilde{\alpha}_2}{\alpha_1}\right)\left(1 + \frac{\tilde{\beta}_2}{\beta_1}\right) \inf_{v_\Lambda\in Y_\Lambda} \|y - v_\Lambda\|_Y + \frac{\tilde{\beta}_2}{\alpha_1} \inf_{q_\Lambda\in Q_\Lambda} \|p - q_\Lambda\|_Q \\
\|p - p_\Lambda\|_Q &\leq \left(1 + \frac{\tilde{\beta}_2}{\beta_1}\right) \inf_{q_\Lambda\in Q_\Lambda} \|p - q_\Lambda\|_Q + \frac{\tilde{\alpha}_2}{\beta_1}\|y - y_\Lambda\|_Y,
\end{aligned} \tag{4.1.30}$$

hold where α_1, β_1 are the constants from the discrete conditions (4.1.25) and (4.1.28).

Thus, the error between the approximation and the solution can be estimated in terms of the best approximation from Y_Λ, respectively Q_Λ, and the constants regulating the continuity and discrete ellipticity of A and B. For uniform refinements, the best approximation properties are reflected by Jackson estimates of the type (3.2.55) by which one can further estimate the right hand sides in (4.1.30).

For the numerical solution of the finite discrete system

$$\mathbf{L}_\Lambda \begin{pmatrix} \mathbf{y}_\Lambda \\ \mathbf{p}_\Lambda \end{pmatrix} := \begin{pmatrix} \mathbf{A}_\Lambda & \mathbf{B}_\Lambda^T \\ \mathbf{B}_\Lambda & \mathbf{0} \end{pmatrix} \begin{pmatrix} \mathbf{y}_\Lambda \\ \mathbf{p}_\Lambda \end{pmatrix} = \begin{pmatrix} \mathbf{f}_\Lambda \\ \mathbf{u}_\Lambda \end{pmatrix} \tag{4.1.31}$$

which is finally of the form (2.36), i.e., $\mathbf{L}_\Lambda$ is an ℓ_2–automorphism, we can employ any convergent iterative method for saddle point problems. Since the system (4.1.31) has already uniformly bounded condition numbers, we can even use the simplest one, namely, the classical incomplete *Uzawa method*, see e.g. [Br, BF]. Here one solves for a sufficiently small parameter $\gamma > 0$ the equations

$$\begin{aligned} \mathbf{y}_\Lambda &\leftarrow (\mathbf{I}_\Lambda - \mathbf{A}_\Lambda)\mathbf{y}_\Lambda + \mathbf{f}_\Lambda - \mathbf{B}_\Lambda^T \mathbf{p}_\Lambda \\ \mathbf{p}_\Lambda &\leftarrow \mathbf{p}_\Lambda - \gamma\, \mathbf{B}_\Lambda \mathbf{y}_\Lambda \end{aligned} \tag{4.1.32}$$

alternately, starting with an initial approximation for $\mathbf{p}_\Lambda$ and $\mathbf{y}_\Lambda$. In fact, it is shown below in Remark 6.26 that the convergence rate of an incomplete Uzawa method of the form (4.1.32) does *not* depend on Λ. The subject of the numerical solution of the systems (4.1.31) will be taken up again in Section 6.6.4.

The construction of iterative methods and preconditioners for saddle point problems has been the object of many investigations, see e.g. [BWY, BP, BPV1]. In the framework of wavelets, the first preconditioner for saddle point problems of the form (4.2.9) below that yield uniformly bounded condition numbers has been constructed in [K1], see also [K3].

4.2 Elliptic Boundary Value Problems as Saddle Point Problems

Now we are ready to formulate elliptic boundary value problems as saddle point problems, placing particular emphasis on a flexible treatment of *essential, non–homogeneous* boundary conditions.

Let $\Omega \subset \mathbb{R}^n$ be the (bounded) domain from Section 3.1 with Lipschitz continuous boundary $\partial\Omega$ and let $\Gamma \subseteq \partial\Omega$ be a smooth subset. The interesting case will be when the surface measure of Γ is strictly positive.

In order to motivate the main ideas, we consider first the elliptic boundary value problem

$$\begin{aligned} -\nabla \cdot (\mathbf{a}\nabla y) + k\,y &= f \qquad \text{in } \Omega, \\ y &= u, \qquad \text{on } \Gamma \\ (\mathbf{a}\nabla y)\cdot \mathbf{n} &= 0, \qquad \text{on } \partial\Omega \setminus \Gamma \end{aligned} \tag{4.2.1}$$

where $\mathbf{n} = \mathbf{n}(\mathbf{x})$ is the outward normal at $\mathbf{x} \in \partial\Omega \setminus \Gamma$. Moreover, $\mathbf{a}(\mathbf{x}) = (a_{i,j}(\mathbf{x}))_{i,j}$ is uniformly positive definite on Ω, and $ky := \mathbf{b} \cdot \nabla y + a_0 y \geq 0$ with $b_i, a_0 \in L_\infty(\Omega)$.

In view of Chapter 6 where u will be a boundary control, the boundary condition will be already denoted by u but will be assumed to be *given* throughout this chapter.

A standard derivation of a weak formulation for (4.2.1) yields the following problem, see e.g. [Ha2]: Given $f \in (H^1(\Omega))'$ and some $u \in H^{1/2}(\Gamma)$, determine the solution $y \in H^1(\Omega)$ satisfying

$$\begin{aligned} a_\Omega(y, v) &= \langle f, v\rangle_\Omega \qquad \text{for all } v \in H^1_{0,\Gamma}(\Omega), \\ y|_\Gamma &= u, \end{aligned} \tag{4.2.2}$$

where $H^1_{0,\Gamma}(\Omega) := \{v \in H^1(\Omega) : v|_\Gamma = 0\}$. By $\langle\cdot,\cdot\rangle_\Omega$ we abbreviate here the corresponding dual form between function spaces living on Ω. Note that the bilinear form $a_\Omega(\cdot,\cdot)$ on $H^1(\Omega) \times H^1(\Omega)$ defined by

$$a_\Omega(v, w) := \int_\Omega (\mathbf{a}\nabla v \cdot \nabla w + k\, vw)\, d\mathbf{x}$$

is continuous and elliptic on $H^1_{0,\Gamma}(\Omega)$. The Neumann boundary conditions on $\partial\Omega \setminus \Gamma$ do not explicitly appear any more in the weak form (4.2.2), hence, they are called *natural* boundary conditions. In contrast, the Dirichlet boundary conditions in (4.2.1) still have to be posed in (4.2.2), inducing the term *essential* boundary conditions.

For homogeneous Dirichlet boundary conditions $u \equiv 0$ on Γ, the boundary conditions in (4.2.2) can be handled by including them also into the solution space for y. Then (4.2.2) reduces to the following form: Given $f \in (H^1_{0,\Gamma}(\Omega))'$, find $y \in H^1_{0,\Gamma}(\Omega)$ such that

$$a_\Omega(y, v) = \langle f, v\rangle_\Omega \quad \text{for all } v \in H^1_{0,\Gamma}(\Omega) \tag{4.2.3}$$

holds. These equations are just the optimality conditions for the quadratic functional

$$\inf_{v \in H^1_{0,\Gamma}(\Omega)} \tfrac{1}{2} a_\Omega(v, v) - \langle f, v\rangle_\Omega. \tag{4.2.4}$$

In order to treat inhomogeneous boundary conditions $y|_\Gamma = u$ in (4.2.3), one typically uses a homogenization technique as follows. One determines a function $y_0 \in H^1(\Omega)$ which satisfies $y_0|_\Gamma = u$. Then one solves (4.2.3) for the modified right hand side $f - a_\Omega(y_0, v)$. This procedure would have to be repeated whenever a change in the boundary function u occurs, like in the case of optimal control problems with boundary control discussed in Chapter 6.

4.2.1 The Fictitious Domain—Lagrange Multiplier Approach

Therefore, one looks for weak formulations where (essential) boundary conditions are *not* included into the solution or approximation space. A common approach to do that is the Lagrange multiplier method. In the present context, it is most advantageous when combined with a fictitious domain. To describe this, let $\square$ be the fictitious domain containing Ω (3.1.1). Denote the extension of $f \in (H^1(\Omega))'$ to $\square$ again by f. Let $a(\cdot,\cdot) := a_\square(\cdot,\cdot)$ be the corresponding continuous 'extension' of $a_\Omega(\cdot,\cdot)$ to $H^1(\square) \times H^1(\square)$,

$$a(v,w) = \int_\square (\mathbf{a}\nabla v \cdot \nabla w + k\, vw)\, d\mathbf{x}, \tag{4.2.5}$$

and let $u \in H^{1/2}(\Gamma)$ be as before a given boundary function.

In view of the abstract setting presented in Section 4.1.1, we define a bilinear form $b(\cdot,\cdot)$ induced by the restriction operator in (4.2.1) as follows. We explicitly express the restriction of a function in $v \in H^1(\square)$ to Γ, usually denoted by $(\cdot)|_\Gamma$, in terms of the *trace operator* B,

$$Bv := v|_\Gamma. \tag{4.2.6}$$

Recall that for any $v \in H^1(\square)$ its trace Bv is known to be in $H^{1/2}(\Gamma)$, see e.g. [Gr] and Section 4.2.2 below. Thus, the bilinear form

$$b(v,q) \;:=\; \langle Bv, q\rangle_{H^{1/2}(\Gamma)\times(H^{1/2}(\Gamma))'} = \langle Bv, q\rangle_\Gamma \tag{4.2.7}$$

is well-defined on $H^1(\square) \times (H^{1/2}(\Gamma))'$.

Instead of (4.2.4), consider now the following *saddle point problem*: For $f \in (H^1(\square))'$ and $u \in H^{1/2}(\Gamma)$, find the solution of

$$\inf_{v\in H^1(\square)} \sup_{q\in(H^{1/2}(\Gamma))'} \tfrac{1}{2}a(v,v) - \langle f, v\rangle_\square + b(v,q) - \langle u, q\rangle_\Gamma. \tag{4.2.8}$$

Thus, instead of enforcing the essential boundary conditions within $H^1(\Omega)$ one embeds Ω into $\square$ and introduces an additional variable q called the *Lagrange multiplier* to *append* the constraints imposed by the boundary conditions to a quadratic functional of the form (4.2.4). Such an approach is often employed in optimal control theory whenever nonlinear functionals have to be minimized subject to constraints.

Remark 4.5 *In the context of second order elliptic boundary value problems, this approach has been introduced for $\Gamma = \partial\Omega$ in [Ba1]. There the formulation is as follows: Given $(f,u) \in (H^1(\Omega))' \times H^{1/2}(\Gamma)$, find $(y,p) \in H^1(\Omega) \times (H^{1/2}(\Gamma))'$ such that*

$$\begin{aligned} a_\Omega(y,v) + b(v,p) &= \langle f, v\rangle_\Omega \qquad \textit{for all } v \in H^1(\Omega), \\ b(y,q) &= \langle u, q\rangle_\Gamma \qquad \textit{for all } q \in (H^{1/2}(\Gamma))' \end{aligned} \tag{4.2.9}$$

holds. This approach is also motivated by the derivation of the weak formulation using Green's formula. In fact, in the special case of the Dirichlet problem the Lagrange multiplier turns out to be the conormal derivative of y at Γ, $p = \mathbf{n}\cdot(\mathbf{a}\nabla y)$, which can be interpreted as the stress vector *on the boundary.*

Setting up the optimality conditions for the saddle point problem (4.2.8) yields the following problem: Given $(f,u) \in (H^1(\square))' \times H^{1/2}(\Gamma)$, find $(y,p) \in H^1(\square) \times (H^{1/2}(\Gamma))'$ such that

$$\begin{aligned} a(y,v) + b(v,p) &= \langle f, v\rangle_\square \qquad \text{for all } v \in H^1(\square), \\ b(y,q) &= \langle u, q\rangle_\Gamma \qquad \text{for all } q \in (H^{1/2}(\Gamma))' \end{aligned} \tag{4.2.10}$$

holds. In order to write (4.2.10) in operator form, let the linear operator A be defined by

$$\langle Av, w\rangle_{(H^1(\square))'\times H^1(\square)} := a(v,w) \tag{4.2.11}$$

and let the dual of B' be given by

$$\langle v, B'q\rangle_{H^1(\square)\times(H^1(\square))'} := b(v,q) = \langle Bv, q\rangle_{H^{1/2}(\Gamma)\times(H^{1/2}(\Gamma))'}. \tag{4.2.12}$$

Then in compact operator form, (4.2.10) is rewritten as:

Fictitious Domain—Lagrange Multiplier Approach: Given $(f,u) \in (H^1(\square))' \times H^{1/2}(\Gamma)$, find $(y,p) \in H^1(\square) \times (H^{1/2}(\Gamma))'$ such that

$$\mathcal{L}\begin{pmatrix} y \\ p \end{pmatrix} := \begin{pmatrix} A & B' \\ B & 0 \end{pmatrix}\begin{pmatrix} y \\ p \end{pmatrix} = \begin{pmatrix} f \\ u \end{pmatrix}. \tag{4.2.13}$$

This combined Fictitious Domain—Lagrange Multiplier Approach has the following advantages:

- The formulation (4.2.10) provides a 'decoupling' of the differential operator from the constraints, the essential boundary conditions;
- as a consequence, both changing boundary conditions and changing boundaries can be treated adequately by modifying only either u or B, respectively; this makes it particularly appropriate for control problems with boundary control or for shape optimization problems;
- entailing that for the numerical solution the main amount of work, namely, the set–up and application of A, can be executed on a very simple domain where fast and very efficient methods exist;
- while the treatment of the boundary only involves a lower dimensional domain.
- It allows for (asymptotically) optimal preconditioning.

However, the gain of information is payed for by the fact that the linear system (4.2.13) is *indefinite* and thus requires different iterative solvers than positive definite systems. The fact that (4.2.13) involves more unknowns is compensated by the observation that the additional variable has an important physical meaning, see [GuH, GHS1].

A few remarks on the relations between the solutions of the original problem in strong form (4.2.1), problem (4.2.9) and the extended formulation (4.2.13) are in order.

Remark 4.6 *For $\Gamma = \partial\Omega$, one can show that the solution $y \in H^1(\square)$ of (4.2.13), when restricted to Ω, solves the saddle point problem (4.2.9) [GPP].*

This is no longer automatically the case when $\Gamma \subset \partial\Omega$. The fact that one has both Dirichlet and Neumann boundary conditions in (4.2.1) poses additional difficulties due to a possible lack of regularity at the interfaces. In particular, the (natural) boundary conditions are no longer implicitly contained in the weak formulation. In the context of bivariate shape optimization problems, this subject is discussed in [KP]. There the *boundary part Γ is sought as the solution of an optimization problem, and extensions of problems of the type (4.2.3) to $\square$ are constructed whose solution again satisfies (4.2.1) when restricted to Ω. For the present situation, this will be investigated elsewhere.*

Both the strong form (4.2.1) as well as the weak form (4.2.9) have been recalled here firstly for motivation, and the natural boundary conditions $((\mathbf{a}\nabla y)\cdot\mathbf{n})|_{\partial\Omega\setminus\Gamma} = 0$ is included in (4.2.1) to pose a well-defined problem.

In Chapter 6, optimal control problems will be considered where the constraints are elliptic boundary value problems in the weak form (4.2.13). For this purpose, it is important that (4.2.13) possesses a unique solution.

In any case, the Lagrange multiplier p solving (4.2.13) coincides with the Lagrange multiplier in (4.2.9) since the boundary inner product $b(\cdot,\cdot)$ is the same for functions in $H^1(\square)$ and $H^1(\Omega)$.

Different fictitious domain approaches have been employed for quite a long time, both on the continuous as well as on the matrix level where this is called *matrix capacitance method*, see for instance [Ag, As, FK, GPP, HHK, KP, MKM, Nep, PW, Ri1]. Lately they have become even more interesting since, for 3D problems, mesh generation of Ω can become quite expensive. Of course, the above mentioned advantages will only prevail if there is no need to adapt the discretization on $\square$ to the particular current domain geometry Ω. This is essential when dealing with moving boundaries. But even for fixed domains this is still very desirable when aligning the mesh with the boundary of Ω causes strong distortions. For this reason, in the discretization procedure the strategy pursued here is to keep the particular form of the discretization of the Lagrange multipliers *completely independent* of the (fixed) trial spaces on $\square$. This also means *not to distort* the different meshes on $\square$ and Γ as it is done in some finite element contexts for stability reasons, see e.g. [Pi]. As it is discussed in Section 4.2.4 below, stability requires some balancing of the difference of *resolution levels* of the discretizations.

One should mention another standard technique for appending essential boundary conditions: to use *penalty parameters* [Ba2]. For flow problems, the penalization approach combined with a fictitious domain technique has been investigated in [An, ABF]. When the essential boundary conditions are enforced by penalization, one appends a term of the form $\varepsilon^{-1}(Bv-u,Bv)_{L_2(\Gamma)}$ (or an inner product $c(\cdot,\cdot)$ corresponding to a smoother norm on Γ) to the functional in (4.2.4). This idea is not further pursued here for the following reason. In order to increase accuracy of the boundary constraints as the refinement level j increases, one chooses the penalty parameter as $\varepsilon \sim 2^{-rj}$. Here $r>0$ is some fixed real parameter depending on the choice of the inner product $c(\cdot,\cdot)$ and chosen such that the standard error estimates are not aggravated. Unfortunately, such a choice of ε causes the condition number of the corresponding differential operator $A_\varepsilon := A+\varepsilon^{-1}B'B$ to grow at least like ε^{-1} [K1]. So far no asymptotically optimal preconditioner for this situation is known. At least from the experiments in [Ri1, Ri2] for a comparable situation, one can conclude that applying asymptotic optimal preconditioners for A like the BPX–preconditioner or a wavelet preconditioner provides some improvement over no preconditioning at all. This can be explained by the fact that A is an operator of order two while $B'B$ is of order one. Thus, a preconditioner tailored to A transforms A_ε only into an operator of order one. In contrast, viewed from the point of preconditioning the Lagrange multiplier approach seems to be better suited: it does allow for optimal preconditioning as detailed in Chapter 2.

Remark 4.7 *In principle, the Fictitious Domain—Lagrange Multiplier Approach is neither confined to the usual trace operator B in (4.2.6) nor to second order problems. In fact, the techniques apply as long as the operator A satisfies the inf–sup conditions (4.1.10) and for the operator B a trace theorem Theorem 4.9 below holds, implying the inf–sup condition (4.1.11), and A and B are continuous.*

This approach can also be used for problems where the operator A itself represents already a saddle point operator, like the Stokes problem where the incompressibility condition appears as an additional constraint, see [GuH, DKS2].

In order to establish Step 1 of the general concept, the well–posedness of problem (4.2.13), we apply Theorem 4.1. To check the conditions (4.1.10) and (4.1.11) and the continuity conditions, note first that the bilinear form $a(\cdot,\cdot)$ defined in (4.2.5) is continuous on $H^1(\Box)\times H^1(\Box)$. Moreover, $a(\cdot,\cdot)$ satisfies the ellipticity conditions (4.1.10) on all of $H^1(\Box)$ if $k > 0$ and therefore in particular on $\ker B$ as required in Theorem 4.1.

Remark 4.8 *If $k \equiv 0$ in (4.2.1), the bilinear form $a(\cdot,\cdot)$ is still elliptic on the kernel of B if Γ has a strictly positive measure, see [Ci1], Theorem 1.2.1.*

In the next section the continuity for $b(\cdot,\cdot)$ and the inf–sup condition (4.1.11), which takes on here the form

$$\inf_{q\in(H^{1/2}(\Gamma))'} \sup_{v\in H^1(\Box)} \frac{b(v,q)}{\|v\|_{H^1(\Box)}\,\|q\|_{(H^{1/2}(\Gamma))'}} \geq \tilde{\beta}_1 > 0 \quad \text{for some } \tilde{\beta}_1 \in \mathbb{R}^+, \tag{4.2.14}$$

will be derived from the classical Trace Theorem. For the present situation, condition (4.2.14) expresses that the trace of functions in $H^1(\Box)$ should not be in the space orthogonal to $(H^{1/2}(\Gamma))'$ with respect to $\langle\cdot,\cdot\rangle_\Gamma$ where the 'amount' of non-orthogonality is quantified by the parameter $\tilde{\beta}_1$.

4.2.2 Trace Theorems

Let us recall the classical trace theorem from e.g. [Gr] for domains Ω with a Lipschitz continuous boundary $\partial\Omega$.

Theorem 4.9 *For any $f \in H^s(\Omega)$, $1/2 < s < 3/2$, one has*

$$\|f|_{\partial\Omega}\|_{H^{s-1/2}(\partial\Omega)} \leq c_{T1}\,\|f\|_{H^s(\Omega)}. \tag{4.2.15}$$

Conversely, for every $h \in H^{s-1/2}(\partial\Omega)$, there exists some $f \in H^s(\Omega)$ such that $f|_{\partial\Omega} = h$ and

$$\|f\|_{H^s(\Omega)} \leq c_{T2,\Omega}\,\|h\|_{H^{s-1/2}(\partial\Omega)}. \tag{4.2.16}$$

Note that the range of s extends accordingly if $\partial\Omega$ is more regular.

Next we extend the Trace Theorem in the context of fictitious domain methods where $\Omega \subseteq \Box$ and for smooth subsets $\Gamma \subset \partial\Omega$. Recall that Whitney–extension results, see e.g. [Gr], Section 1.4, ensure that any function $f \in H^s(\Omega)$ can be extended to $\hat{f} \in H^s(I\!R^n)$ such that $\hat{f}|_\Omega = f$ and

$$\|\hat{f}\|_{H^s(I\!R^n)} \leq c_E \|f\|_{H^s(\Omega)}, \qquad s > 0. \tag{4.2.17}$$

Since $\|\hat{f}\|_{H^s(\Box)} \leq \|\hat{f}\|_{H^s(I\!R^n)}$, it follows that

$$\|\hat{f}\|_{H^s(\Box)} \leq c_E \|f\|_{H^s(\Omega)}, \qquad s > 0. \tag{4.2.18}$$

By the same argument, for any $h \in H^{s-1/2}(\Gamma)$, there exists an extension $\hat{h} \in H^{s-1/2}(\partial\Omega)$ such that

$$\|\hat{h}\|_{H^{s-1/2}(\partial\Omega)} \leq c_{\partial\Omega} \|h\|_{H^{s-1/2}(\Gamma)}, \qquad s > 0. \tag{4.2.19}$$

Thus, an immediate consequence of Theorem 4.9 and (4.2.18), (4.2.19) is the following result.

Corollary 4.10 *For any $f \in H^s(\Box)$, $1/2 < s < 3/2$, one can estimate*

$$\|Bf\|_{H^{s-1/2}(\Gamma)} \leq c_{T_1} \|f\|_{H^s(\Box)} \tag{4.2.20}$$

with the same constant c_{T1} from (4.2.15).
Conversely, for every $h \in H^{s-1/2}(\Gamma)$, there exists some $f \in H^s(\Box)$ such that $Bf = h$ and

$$\|f\|_{H^s(\Box)} \leq c_{T_2} \|h\|_{H^{s-1/2}(\Gamma)} \tag{4.2.21}$$

where

$$c_{T_2} := c_E \, c_{T_2,\Omega} \, c_{\partial\Omega}, \tag{4.2.22}$$

with c_E from (4.2.18).

Proof: The first part is trivial since for every $f \in H^s(\Box)$ its restriction to Ω, $\bar{f} = f|_\Omega$, satisfies (4.2.15), and for $\Gamma \subseteq \partial\Omega$

$$\|Bf\|_{H^{s-1/2}(\Gamma)} \leq \|f|_{\partial\Omega}\|_{H^{s-1/2}(\partial\Omega)} \leq c_{T_1} \|\bar{f}\|_{H^s(\Omega)} \leq c_{T_1} \|f\|_{H^s(\Box)}.$$

For the second estimate (4.2.21), pick any $h \in H^{s-1/2}(\Gamma)$. Then there is some extension $\hat{h} \in H^{s-1/2}(\partial\Omega)$ satisfying (4.2.19). Applying Theorem 4.9, there exists an $f \in H^s(\Omega)$ such that $Bf = \hat{h}$ satisfying (4.2.16). Denoting by $\hat{f} \in H^s(\Box)$ its extension to $\Box$, we have $B\hat{f} = \hat{h}$ and, by (4.2.18) and (4.2.16),

$$\|\hat{f}\|_{H^s(\Box)} \leq c_E \|f\|_{H^s(\Omega)} \leq c_E c_{T2,\Omega} \|\hat{h}\|_{H^{s-1/2}(\partial\Omega)} \leq c_{T_2} \|h\|_{H^{s-1/2}(\Gamma)}. \tag{4.2.23}$$

■

From this, it is rather trivial to conclude the inf–sup condition (4.2.14).

Remark 4.11 *For $b(\cdot,\cdot)$ defined by (4.2.7) on $H^1(\Box)\times(H^{1/2}(\Gamma))'$ and c_{T_2} from (4.2.22), one can infer the inf-sup condition with respect to $H^1(\Box)\times(H^{1/2}(\Gamma))'$,*

$$\inf_{q\in(H^{1/2}(\Gamma))'}\ \sup_{v\in H^1(\Box)} \frac{b(v,q)}{\|v\|_{H^1(\Box)}\,\|q\|_{(H^{1/2}(\Gamma))'}} \geq \frac{1}{c_{T_2}}. \tag{4.2.24}$$

Proof: By definition of the norm, one has for every $q\in(H^{1/2}(\Gamma))'$

$$\begin{aligned}\|q\|_{(H^{1/2}(\Gamma))'} &= \sup_{h\in H^{1/2}(\Gamma)} \frac{\langle q,h\rangle_{(H^{1/2}(\Gamma))'\times H^{1/2}(\Gamma)}}{\|h\|_{H^{1/2}(\Gamma)}}\\ &= \sup_{f\in H^1(\Box),\ Bf\in H^{1/2}(\Gamma)} \frac{\langle q,Bf\rangle_{(H^{1/2}(\Gamma))'\times H^{1/2}(\Gamma)}}{\|Bf\|_{H^{1/2}(\Gamma)}}\end{aligned}$$

since, by the second part of the Trace Theorem 4.10 there exists for every $h\in H^{1/2}(\Gamma)$ an $f\in H^1(\Box)$ such that $h=Bf$. Applying (4.2.21) while realizing that taking the supremum over all $v\in H^1(\Box)$ means taking the supremum over a bigger set yields (4.2.7). ∎

4.2.3 Conditions Ensuring Stability of the Discretizations: General Remarks

Next we specify the situation considered in Section 4.1.3 to the elliptic boundary value problem (4.2.10). Let

$$Y_\Lambda\subset Y=H^1(\Box)) \quad\text{and } Q_\Lambda\subset Q=L_2(\Gamma)\subset(H^{1/2}(\Gamma))')$$

be finite–dimensional subspaces where again Λ consists of two index sets Λ_Y,Λ_Q which are (at first) independent of each other.

For the situation at hand, we need to derive conditions that ensure that Y_Λ, Q_Λ provide stable discretizations, formulated in terms of the discrete ellipticity conditions (4.1.25) and the LBB condition (4.1.28).

The bilinear form $a(\cdot,\cdot)$ defined in (4.2.5) is for $k>0$ elliptic on all of $Y=H^1(\Box)$ and, thus, on any finite–dimensional subspace of Y. This is the problem that we will consider in the sequel. For $k=0$ one would have to check the discrete ellipticity condition (4.1.25) on the kernel of B_Λ for the particular case at hand.

Thus, we concentrate on ensuring (4.1.28). Recall from Remark 4.3 that the strong condition (4.1.29) would imply that the discrete space $\tilde{Q}_\Lambda\subset H^{1/2}(\Gamma)$ is determined by the *traces* of functions in Y_Λ. One situation where this applies is when the faces of the boundary Γ are parallel to Cartesian coordinate axes and when the generators on the domain are tensor products of univariate functions. For general domains which are treated in combination with a fictitious domain approach, condition (4.1.29) would induce irregular grids on Γ even when the grid on $\Box$ is uniform, making it difficult to construct wavelets on Γ.

4.2.4 Conditions Ensuring Stability of the Discretizations: The LBB Condition

The results in this subsection are essentially contained in [DK2]. We treat here the case of uniform refinements and adhere again to the notation of Section 3.2. The generators and wavelets are indexed in addition by $\Box$ and Γ standing for the domain or boundary on which they live. That this, let there be two multiresolution analyses $\mathcal{Y}$ of $H^1(\Box)$ and $\mathcal{Q}$ of $L_2(\Gamma)$ where the discrete spaces are specialized to

$$Y_\Lambda =: Y_j \subset H^1(\Box) \quad \text{and} \quad Q_\Lambda =: Q_\ell \subset L_2(\Gamma) \subset (H^{1/2}(\Gamma))'$$

(so that $\langle \cdot, \cdot \rangle_\Gamma$ can be identified with $(\cdot, \cdot)_{L_2(\Gamma)}$). The spaces are represented by

$$\begin{aligned} Y_j &= S(\Phi_j^\Box) &= S(\Psi^{j,\Box}), \qquad \tilde{Y}_j &= S(\tilde{\Phi}_j^\Box) &= S(\tilde{\Psi}^{j,\Box}), \\ Q_\ell &= S(\Phi_\ell^\Gamma) &= S(\Psi^{\ell,\Gamma}), \qquad \tilde{Q}_\ell &= S(\tilde{\Phi}_\ell^\Gamma) &= S(\tilde{\Psi}^{\ell,\Gamma}). \end{aligned} \tag{4.2.25}$$

Here the indices j and ℓ refer to mesh sizes on the domain and the boundary,

$$h_\Box \sim 2^{-j} \qquad \text{and} \qquad h_\Gamma \sim 2^{-\ell}.$$

The discrete inf–sup condition (4.1.28), the LBB condition, for the pair Y_j, Q_ℓ requires that there exists a constant $\beta_1 > 0$ *independent* of j and ℓ such that

$$\inf_{q\in Q_\ell} \sup_{v\in Y_j} \frac{b(v,q)}{\|v\|_{H^1(\Box)} \, \|q\|_{(H^{1/2}(\Gamma))'}} \;\geq\; \beta_1 > 0 \tag{4.2.26}$$

holds. Recall that we are particularly interested in such situations where the Q_ℓ are *not* trace spaces of Y_j, i.e., where a condition of the form (4.1.29) is not satisfied.

It has been observed already in [Ba1] that (4.2.26) constrains the choice of $h_\Box$ and h_Γ. In fact, (for $H^1(\Omega)$ and $(H^{1/2}(\Gamma))'$) the LBB condition is satisfied for general situations whenever

$$\frac{h_\Gamma}{h_\Omega} = 2^{j-\ell} \geq c_\Omega > 1 \tag{4.2.27}$$

holds for some sufficiently large constant c_Ω which depends on the domain in a complicated way. Thus, the mesh size on the boundary is required to be *larger* than that on the domain. In contrast, intuitively one would think that the smaller the mesh size for the Lagrange multipliers is chosen, the more accurately one can enforce the boundary constraints. On the other hand, standard a–priori error estimates like (4.1.16) involve β_1 from (4.2.26) in the denominator. Thus, one faces interfering demands.

To investigate situations and derive conditions for the validity of the LBB condition has been the object of several investigations, besides the first article [Ba1]. Under an assumption of the type (4.2.27), the results in [Br1] focus on improving error estimates and on numerically solving the resulting discrete finite–dimensional problem. Other studies attempt to identify possibly small constants c_Ω in (4.2.27) that still imply (4.2.26). All these contributions may be roughly divided into two groups. In one group the discretization of the Lagrange multipliers is closely tied to the trial spaces on the domain, e.g., using traces of trial functions when the boundary is aligned with mesh lines, see e.g. [Pi]. In the second group represented by [GG, GPP], the position of the boundary Γ is less constrained. Nevertheless, in both settings only *bivariate* problems are treated, and Γ is typically assumed to be (or to be approximated by) a polygon. Moreover, in these investigations the spaces Y_j and Q_ℓ consist only of piecewise linear and piecewise constant functions, respectively. In this case, it has been shown that the LBB condition is satisfied for $c_\Omega \approx 2$ or 3 in (4.2.27).

It should be mentioned already at this point that the obstructions caused by the LBB condition can be avoided by means of a stabilization technique proposed in [St1] where, however, the location of the boundary of Ω relative to the mesh is somewhat constrained. Another stabilization strategy based on wavelets has been investigated in [Be3]. A related approach which systematically avoids restrictions of the LBB type is based on least squares techniques and is described in [DKS2], see also Section 5.

Objectives: In view of the above mentioned applications and, in particular, Section 6, the aim now is to derive conditions for the validity of the LBB condition under the following premises:

- The spatial dimension of Ω is *arbitrary.*
- The requirements on the trial spaces concern only *locality* and *approximation order.* In particular, all classical finite element spaces as well as multiresolution settings like in Section 3.2 are covered.
- The boundary Γ of Ω is *independent* of the stationary hierarchy of meshes for the fixed domain $\Box$.
- The Lagrange multiplier spaces Q_ℓ are *independent* of the trial spaces on Ω or $\Box$. In particular, they are *not* assumed to be traces of Y_j.

Of course, in such a framework one cannot expect to obtain concrete values for c_Ω in (4.2.27). Instead, we are rather interested in bringing out *which parameters* of a given setting determine the size of c_Ω, or better $c_\Box$. Precisely, it will be shown how β_1 in (4.2.26) depends under most general assumptions on the discretizations on the different constants. These are the ones appearing in the Trace Theorem and in norm equivalences for the spaces $H^{1/2}(\Gamma)$ and $(H^{1/2}(\Gamma))'$, evaluated in terms of wavelets.

In contrast to [GG], the approach used here does not employ Fortin's condition [BF]. It also differs from [Ba1] in several respects and produces slightly different results. Neither is H^2–regularity of the solution of (4.2.2) required here nor do the Lagrange multipliers have to belong to $H^{1/2}(\Gamma)$ which would exclude piecewise constants. Moreover, in contrast to previous studies the idea here is that the traces of Y_j should be able to approximate the elements of a suitable system *dual* to the Lagrange multipliers possibly well. In particular, this dual system could be much more regular than the Lagrange multipliers. The quantification of this concept differs from [Ba1] and proceeds in the following two steps. A first crucial ingredient is the sharp Cauchy–Schwarz inequality from Lemma 3.12 which will then be combined with a perturbation argument. Specializing the setting to the case that wavelet bases on $\Box$ and Γ are available, this reasoning leads to criteria which involve only a *one parameter family* for a *single* constant. In fact, it relates the LBB condition to the *spectral properties* of *wavelet representations* of the trace operator. This is an observation which might be of interest in its own right since it can be used to numerically compute estimates for the LBB constants in very general situations. Due to the option of choosing *different* dual systems for the Lagrange multiplier spaces, this allows to exploit higher regularity of the boundary Γ independently of the regularity of the Lagrange multipliers.

Remark 4.12 *In order to derive conditions that guarantee that the LBB condition (4.2.26) holds under most general circumstances, one does* not *need to assume from the beginning that wavelet bases for $H^1(\Box)$ and $(H^{1/2}(\Gamma))'$ are available having all the properties listed in Section 3.2. In fact, the first results in [DK2] are established under the following* assumptions *making* no *explicit use of wavelet bases:*

(A1) L_2-stability of $\mathcal{Q}, \tilde{\mathcal{Q}}$:
$\Phi_j^\Gamma, \tilde{\Phi}_j^\Gamma$ *generating the spaces $Q_j, \tilde{Q}_j$ satisfy the biorthogonality conditions (3.2.35) or, weaker, for $Q_j, \tilde{Q}_j$ the reverse Cauchy–Schwarz inequality (3.2.54) holds;*

(A2) Jackson and Bernstein estimates for $\tilde{\mathcal{Q}}$:
for $\tilde{\mathcal{Q}}$ a direct estimate (3.2.55) holds up to order $\tilde{d}_\Gamma$, $\tilde{d}_\Gamma$ a fixed integer , i.e.,

$$\inf_{\tilde{q}_j\in\tilde{Q}_j} \|q - \tilde{q}_j\|_{L_2} \le c_\Gamma\, 2^{-j\tilde{d}_\Gamma} \|q\|_{H^{\tilde{d}_\Gamma}(\Gamma)}, \quad q \in H^{\tilde{d}_\Gamma}(\Gamma), \tag{4.2.28}$$

and the inverse inequality

$$\|\tilde{q}\|_{H^s(\Gamma)} \le c_{B\Gamma}\, 2^{js} \|\tilde{q}\|_{L_2(\Gamma)}, \quad \tilde{q} \in \tilde{Q}_j,\ s < \tilde{t}_\Gamma, \tag{4.2.29}$$

is satisfied for some $\tilde{t}_\Gamma > 1/2$;

(A3) Jackson estimates for $\mathcal{Y}$:
for $1 < s \le d_\Box$, $d_\Box$ a positive integer, the Jackson estimates

$$\inf_{v_j\in Y_j} \|v - v_j\|_{H^1(\Box)} \le c_{J\Box}\, 2^{-j(s-1)} \|v\|_{H^s(\Box)}, \quad v \in H^s(\Box), \tag{4.2.30}$$

hold.

The latter relation describes the approximation rates achieved by the spaces Y_j. Its biggest number $d_\Box$ is typically the order *of the generators. More generally, it is well-known that (4.2.30) holds whenever the Y_j are spanned by stable local bases (in the sense of (3.2.4)) and contain all polynomials up to the order $d_\Box$.*

Here we will focus on those aspects which arise when wavelet bases are available for the involved spaces as described initially. Recall from Section 3.2.2 that the validity of the inequalities (4.2.28) and (4.2.29) does indeed provide the type of norm equivalences needed below.

The main theorem, Theorem 4.15 below, will be formulated in terms of the discrete norms (3.2.66) expressed by means of the wavelet bases $\Psi^{j,\square}$, $\Psi^{\ell,\Gamma}$. An additional index refers to the domain or manifold, i.e.,

$$|q|_{s,\Gamma} := \left(\sum_{j=j_0-1}^{\infty} 2^{2sj} \|\langle q, \tilde{\Psi}_j^{\Gamma}\rangle^T\|^2_{\ell_2(\nabla_j^{\Gamma})}\right)^{1/2}, \quad |v|_{s,\square} := \left(\sum_{j=j_0-1}^{\infty} 2^{2sj} \|\langle v, \tilde{\Psi}_j^{\square}\rangle^T\|^2_{\ell_2(\nabla_j^{\square})}\right)^{1/2}. \tag{4.2.31}$$

Recall from Corollary 3.7 the norm equivalences

$$|v|_{s,\square} \sim \|v\|_{H^s(\square)}, \quad s \in (-\tilde{\sigma}_\square, \sigma_\square), \tag{4.2.32}$$

and

$$|q|_{s,\Gamma} \sim \|q\|_{H^s(\Gamma)}, \quad s \in (-\tilde{\sigma}_\Gamma, \sigma_\Gamma), \tag{4.2.33}$$

where $\sigma_\square, \tilde{\sigma}_\square, \sigma_\Gamma, \tilde{\sigma}_\Gamma$ are defined according to (3.2.58). Note that the quantities $|\cdot|_{s,\square}$, $|\cdot|_{s,\Gamma}$ are defined for *any* s while the equivalences (4.2.32), (4.2.33) are confined to the respective ranges for s. The lower estimate in (4.2.32) actually holds for a larger range, namely,

$$|v|_{s,\square} \lesssim \|v\|_{H^s(\square)}, \quad s \le d_\square, \tag{4.2.34}$$

see [DSt]. Thus, setting

$$A^s(\square) := \{v \in L_2(\square) : |v|_{s,\square} < \infty\},$$

(4.2.32) states that $A^s(\square)$ and $H^s(\square)$ agree as sets for $0 < s < \sigma_\square$ and have equivalent norms. Furthermore, define the following factor norms

$$N_s(\tilde{q}) := \inf_{w \in A^{s+1/2}(\square),\ Bw = \tilde{q}} |w|_{s+1/2,\square}. \tag{4.2.35}$$

Recall from Corollary 4.10 that one can find for each $\tilde{q} \in H^s(\Gamma)$ some $w \in H^{s+1/2}(\square)$ such that $Bw = \tilde{q}$ and $\|w\|_{H^{s+1/2}(\square)} \le c_{T_2}\|\tilde{q}\|_{H^s(\Gamma)}$. Thus, one concludes now from (4.2.32) and (4.2.34) the following facts.

Remark 4.13 *For each $0 < s < \tilde{t}_\Gamma, t_\square - 1/2$ there exist positive finite constants $c_{s,\Gamma}, C_{s,\Gamma}$ such that*

$$c_{s,\Gamma} N_s(\tilde{q}) \le |\tilde{q}|_{s,\Gamma} \le C_{s,\Gamma} N_s(\tilde{q}). \tag{4.2.36}$$

Moreover, one has

$$c_{s,\Gamma} N_s(\tilde{q}) \le |\tilde{q}|_{s,\Gamma}, \quad s < \tilde{t}_\Gamma,\ s \le d_\square - 1/2. \tag{4.2.37}$$

Recall from Lemma 3.9 that in the discrete norms $|\cdot|_{s,\square}$, $|\cdot|_{s,\Gamma}$ Jackson and Bernstein estimates hold with constants equal to 1. We will also need the sharp form of the Cauchy Schwarz inequality (3.2.73), namely, that for ever $q \in Q_\ell$ there exists some $\tilde{q}^* \in \tilde{Q}_\ell$ such that

$$|q|_{-1/2,\Gamma} |\tilde{q}^*|_{1/2,\Gamma} = (q, \tilde{q}^*)_{L_2(\Gamma)}. \tag{4.2.38}$$

The first lemma quantifies in these discrete norms how well functions in $\tilde{Q}_\ell$ can approximate traces of functions from Y_j.

Lemma 4.14 *For any fixed $\delta > 0$ satisfying $\delta < \tilde{t}_\Gamma - \frac{1}{2}$ and $\delta \leq d_\Box - 1$, there exists for every $\tilde{q} \in \tilde{Q}_\ell$ some $v_j^* = v_j^*(\tilde{q}) \in Y_j$ such that*

$$|\tilde{q} - Bv_j^*|_{1/2,\Gamma} \;\leq\; K_\delta 2^{-(j+1-\ell)\delta} |\tilde{q}|_{1/2,\Gamma}, \tag{4.2.39}$$

where

$$K_\delta := \frac{C_{1/2,\Gamma}}{c_{\delta+1/2,\Gamma}} \tag{4.2.40}$$

and $c_{s,\Gamma}, C_{s,\Gamma}$ are the constants from (4.2.36). Furthermore, one has for some $0 < \eta < \infty$ the estimate

$$|v_j^*|_{1,\Box} \;\leq\; \left(\frac{\eta}{c_{1/2,\Gamma}} + \frac{2^{-(j+1-\ell)\delta}}{c_{\delta+1/2,\Gamma}}\right) |\tilde{q}^*|_{1/2,\Gamma}. \tag{4.2.41}$$

Proof: Since the unit ball in $A^s(\Box)$ is compact in $L_2(\Box)$, the infimum in (4.2.35) must be attained by some $v \in A^{s+1/2}(\Box)$. Let for each s in the above range $E_s\tilde{q}$ denote the extension of $\tilde{q} \in A^s(\Gamma)$ to $H^{s+1/2}(\Box)$ satisfying

$$N_s(\tilde{q}) \;=\; |E_s\tilde{q}|_{s+1/2,\Box}. \tag{4.2.42}$$

Defining now $v^* := E_{\delta+1/2}\tilde{q}$ and $v_j^* := P_j^\Box v^*$, we first get

$$\begin{aligned} N_{1/2}(\tilde{q} - Bv_j^*) \;&=\; \inf_{w \in H^1(\Box),\; Bw = \tilde{q} - Bv_j^*} |w|_{1,\Box} \\ &\leq\; |v^* - P_j^\Box v^*|_{1,\Box}. \end{aligned}$$

Using (3.2.70), we can conclude from this

$$\begin{aligned} N_{1/2}(\tilde{q} - Bv_j^*) \;&\leq\; 2^{-(j+1)\delta} |E_{\delta+1/2}\tilde{q}|_{1+\delta,\Box} \\ &=\; 2^{-(j+1)\delta}\, N_{\delta+1/2}(\tilde{q}), \end{aligned}$$

which can in turn be estimated by (4.2.37) and (3.2.71) as

$$\begin{aligned} N_{1/2}(\tilde{q} - Bv_j^*) \;&\leq\; \frac{2^{-(j+1)\delta}}{c_{\delta+1/2,\Gamma}} |\tilde{q}|_{\delta+1/2,\Gamma} \\ &\leq\; \frac{2^{-(j+1-\ell)\delta}}{c_{\delta+1/2,\Gamma}} |\tilde{q}|_{1/2,\Gamma}. \end{aligned} \tag{4.2.43}$$

Thus, (4.2.39) follows again by the norm equivalence (4.2.36). The proof of the remaining estimate (4.2.41) for some finite positive constant η follows by the same arguments. ∎

The main theorem for establishing a sufficient condition for the validity of the LBB condition is the following.

Theorem 4.15 *Let the assumptions in Theorem 3.6 be valid such that the norm equivalence (3.2.61) holds for $H^s(\Box)$ for $s \in (-\tilde{\sigma}_\Box, \sigma_\Box)$ with $\sigma_\Box, \tilde{\sigma}_\Box$ defined according to (3.2.58). Suppose that the Trace Theorem Corollary 4.10 holds for $s \leq \tilde{t}_\Gamma, d_\Box - 1$. If, for any refinement level ℓ on Γ, the refinement level j on $\Box$ is chosen as*

$$j \geq \ell + L - 1 \tag{4.2.44}$$

where L is the smallest integer satisfying

$$L > \frac{\log_2 K_\delta}{\delta} \tag{4.2.45}$$

and K_δ is defined by (4.2.40), then the LBB condition (4.2.26), written in terms of the discrete norms as

$$\inf_{q \in Q_\ell} \sup_{v \in Y_j} \frac{b(v,q)}{|v|_{1,\square} |q|_{-1/2,\Gamma}} \geq \beta_1 > 0 \tag{4.2.46}$$

holds with β_1 uniformly bounded away from 0 as $\ell \to \infty$.

Theorem 4.15 actually provides a whole family of estimates for a sufficient large difference of levels depending on the parameter δ. Its range is limited by the approximation order of the trial spaces Y_j and by the regularity of the chosen auxiliary *dual* multiresolution sequence $\tilde{\mathcal{Q}}$.

We have now collected all the ingredients for proving Theorem 4.15. The proof is based on the sharp Cauchy–Schwarz inequality for the spaces on the boundary combined with a perturbation argument involving Lemma 4.14.

Proof of Theorem 4.15: Fix an arbitrary $q \in Q_\ell$. By Lemma 3.12, there exists some $\tilde{q}^* = \tilde{q}^*(q) \in \tilde{Q}_\ell$ such that

$$\begin{aligned} |q|_{-1/2,\Gamma} |\tilde{q}^*|_{1/2,\Gamma} &= (q, \tilde{q}^*)_{L_2(\Gamma)} \\ &= (q, \tilde{q}^* - Bv)_{L_2(\Gamma)} + (q, Bv)_{L_2(\Gamma)} \\ &\leq |q|_{-1/2,\Gamma} |\tilde{q}^* - Bv|_{1/2,\Gamma} + (q, Bv)_{L_2(\Gamma)} \end{aligned}$$

for any $v \in Y_j$. Pick v to be the function $v^* = v^*(\tilde{q}^*) \in Y_j$ that satisfies (4.2.39) and (4.2.41). Then by (4.2.39) one can estimate

$$|q|_{-1/2,\Gamma} |\tilde{q}^*|_{1/2,\Gamma} \leq K_\delta 2^{-(j+1-\ell)\delta} |q|_{-1/2,\Gamma} |\tilde{q}^*|_{1/2,\Gamma} + (q, Bv^*)_{L_2(\Gamma)}$$

where K_δ is given by (4.2.40). Subtraction yields (4.2.41),

$$\frac{(q, Bv_j^*)_{L_2(\Gamma)}}{|q|_{-1/2,\Gamma} \, |\tilde{q}^*|_{1/2,\Gamma}} \geq 1 - K_\delta 2^{-(j+1-\ell)\delta}.$$

from which by (4.2.41) the estimate

$$(q, Bv_j^*)_{L_2(\Gamma)} \geq \left(1 - K_\delta 2^{-(j+1-\ell)\delta}\right) \left(\frac{\eta}{c_{1/2,\Gamma}} + \frac{2^{-(j+1-\ell)\delta}}{c_{\delta+1/2,\Gamma}} \right)^{-1} |q|_{-1/2,\Gamma} \, |v_j^*|_{1,\square} \tag{4.2.47}$$

follows. Thus, whenever j satisfies (4.2.44) with L determined by (4.2.45), one always has that (4.2.46) is satisfied with a constant independent of j, ℓ. Note that the constant on the right hand side of (4.2.41) affects only the *size* of the LBB constant. The *validity* of the LBB condition is already ensured by the right hand side of (4.2.41) which is bounded from below by some positive number. This completes the proof. ∎

Some comments on Theorem 4.15 are in order. The result is similar in spirit with the one established in [Ba1]. Both have in common that a sufficient difference of mesh sizes ensures the validity of the LBB condition without any assumptions concerning a particular interrelation between the trial spaces Y_j on $\square$ and the Lagrange multiplier spaces Q_ℓ. There are, however a few noteworthy distinctions. Aside from the constants from the trace theorem, the involved further constants only reflect properties of the spaces $\tilde{Q}_\ell, Q_\ell$ and Y_j stated in terms of norm equivalences. In contrast to [Ba1], no particular regularity requirements on the solution of (4.2.10) or on the Lagrange multipliers are needed. Moreover, the role of $\tilde{Q}_\ell$ provides additional flexibility resulting in a *family* of estimates.

Here the primary interest is a better understanding of the interplay between discretizations on the domain and on the boundary so that the main value of Theorem 4.15 lies perhaps in the way it is proved. Under the assumptions listed in Remark 4.12, the above result has first been proved in [DK2] employing the norms (3.2.65) defined in terms of the projectors. The resulting estimates are important for the cases where wavelet bases are *not* known explicitly. However, in this situation, switching between the discrete and the function space norms introduces additional constants. This is avoided when one can work as above directly in terms of the norms expressed by means of the wavelets (4.2.31).

In order to gain a more quantitative understanding of the constant K_δ, our next step is to determine the extension $E_s\tilde{q}$ from (4.2.42) in wavelet coordinates. To this end, note that the wavelet representation of the trace of any

$$v = \sum_{j=j_0-1}^{\infty} \sum_{k\in\nabla_j^\square} d_{j,k}\psi_{j,k}^\square \in H^1(\square) =: \mathbf{d}^T\Psi^\square$$

is given by

$$Bv = \sum_{j'=j_0-1}^{\infty} \sum_{k'\in\nabla_{j'}^\Gamma} \left(\sum_{j,k\in\nabla^{\square\cap\Gamma}} d_{j,k}(B\psi_{j,k}^\square, \psi_{j',k'}^\Gamma)_{L_2(\Gamma)} \right) \tilde{\psi}_{j',k'}^\Gamma, \tag{4.2.48}$$

where $\nabla^{\square\cap\Gamma} := \{(j,k) : \operatorname{supp}\psi_{j,k}^\square \cap \Gamma \neq \emptyset\}$. Using the shorthand notation from Section 3.2, (4.2.48) takes on the form

$$Bv = \mathbf{d}^T(B\Psi^\square, \Psi^\Gamma)_{L_2(\Gamma)}\tilde{\Psi}^\Gamma.$$

Let $\mathbf{D}_{s,\square}$, $\mathbf{D}_{s,\Gamma}$ be the (infinite) diagonal matrices defined by

$$(\mathbf{D}_{s,\square})_{(j,k),(j',k')},\ (\mathbf{D}_{s,\Gamma})_{(j,k),(j',k')} := 2^{sj}\delta_{(j,k),(j',k')} \tag{4.2.49}$$

for $(j,k),(j',k') \in \nabla^{\square\cap\Gamma}$, respectively, $(j,k),(j',k') \in \nabla^\Gamma$. In this notation, the norm equivalence (3.2.61) for functions on $\square$ means that the scaled wavelet basis $\mathbf{D}_{-s,\square}\Psi^\square$ is a Riesz basis in $H^s(\square)$. Thus, the function $w^* = \mathbf{d}_*^T\mathbf{D}_{-(s+1/2),\square}\Psi^\square$ solves for a given $\tilde{q} = \tilde{\mathbf{q}}^T\tilde{\Psi}^\Gamma$ the minimization problem

$$\|w^*\|_{s+1/2,\square} = \min_{w\in H^{s+1/2}(\square),\ Bw=\tilde{q}} \|w\|_{s+1/2,\square} = N_s(\tilde{q}) \tag{4.2.50}$$

if and only if $\mathbf{d}_* \in \ell_2(\nabla^{\square\cap\Gamma})$ solves

$$\|\mathbf{d}_*\|_{\ell_2} = \min_{\mathbf{d}' \in \ell_2(\nabla^{\square\cap\Gamma})} \|\mathbf{d}'\|_{\ell_2} \quad \text{subject to} \quad \mathbf{C}_s \mathbf{d}' = \tilde{\mathbf{q}}. \tag{4.2.51}$$

Here we have used the abbreviations

$$\mathbf{C}_s := (\Psi^\Gamma, B\Psi^\square)_{L_2(\Gamma)} \, \mathbf{D}_{-(s+1/2),\square}. \tag{4.2.52}$$

Solving the minimization problem (4.2.51) by a standard Lagrange multiplier approach yields that $\mathbf{d}_*$ is determined by

$$\mathbf{d}_* = \mathbf{C}_s^T \left(\mathbf{C}_s \mathbf{C}_s^T\right)^{-1} \tilde{\mathbf{q}}. \tag{4.2.53}$$

Hence, by definition of the discrete norms (4.2.31), the current version of relation (4.2.36) is equivalent to

$$c_{s,\Gamma} \|\mathbf{d}_*\|_{\ell_2} \;\leq\; \|\mathbf{D}_{s,\Gamma} \tilde{\mathbf{q}}\|_{\ell_2} \;\leq\; C_{s,\Gamma} \|\mathbf{d}_*\|_{\ell_2}.$$

Substituting (4.2.53) and setting $\mathbf{p} := \left(\mathbf{C}_s \mathbf{C}_s^T\right)^{-1} \tilde{\mathbf{q}}$, this becomes

$$c_{s,\Gamma} \|\mathbf{C}_s^T \mathbf{p}\|_{\ell_2} \;\leq\; \|\mathbf{D}_{s,\Gamma} \mathbf{C}_s \mathbf{C}_s^T \mathbf{p}\|_{\ell_2} \;\leq\; C_{s,\Gamma} \|\mathbf{C}_s^T \mathbf{p}\|_{\ell_2}.$$

From this latter relation, one readily concludes that the constants can be identified as the *singular values* of $\mathbf{B}_s$,

$$C_{s,\Gamma} = \sigma_{\max}(\mathbf{B}_s), \quad c_{s,\Gamma} = \sigma_{\min}(\mathbf{B}_s), \tag{4.2.54}$$

where

$$\mathbf{B}_s := \mathbf{D}_{s,\Gamma} \mathbf{C}_s \;=\; \mathbf{D}_{s,\Gamma} \, (\Psi^\Gamma, B\Psi^\square)_{L_2(\Gamma)} \, \mathbf{D}_{-(s+1/2),\square}. \tag{4.2.55}$$

Here $\sigma_{\max/\min}(\mathbf{B}_s) := \sqrt{\lambda_{\max/\min}(\mathbf{B}_s \mathbf{B}_s^T)}$, and $\lambda_{\max/\min}(\mathbf{M})$ denotes the maximal, respectively minimal eigenvalue of the matrix $\mathbf{M}$.

Note that one expects $C_{s,\Gamma}$ to grow and $c_{s,\Gamma}$ to decrease when increasing s. Thus, K_δ should be bounded by the spectral condition number of the *trace matrix* $\mathbf{B}_{\delta+1/2}$ computed as

$$\kappa(\mathbf{B}_{\delta+1/2}) = \frac{\sigma_{\max}(\mathbf{B}_{\delta+1/2})}{\sigma_{\min}(\mathbf{B}_{\delta+1/2})}.$$

In view of Remark 4.11, this appears to be a natural condition.

The present approach differs from all previous ones by the role of the dual multiresolution spaces $\tilde{Q}_\ell$. Although the Lagrange multipliers in Q_ℓ could even be discontinuous, the dual system can be chosen as regular as permitted by the regularity of Γ. A correspondingly high approximation order of the trial spaces on $\square$ would allow to increase the value of δ in (4.2.45). Of course, the spectral condition number of the corresponding trace matrices is expected to grow with increasing δ. However, this growth is damped by the logarithm in the numerator of the right hand side of (4.2.45) while the growth of δ enters the denominator directly. This effect will be confirmed by the numerical experiments reported below.

Remark 4.16 *The trace matrices $\mathbf{B}_{\delta+1/2}$ involve only those wavelets on $\square$ whose supports overlap the boundary. Consequently, the mesh size on $\square$ away from the boundary is not affected by conditions guaranteeing the LBB condition to hold.*

In fact, taking only those wavelets overlapping Γ on a sufficiently fine resolution level near the boundary and grading the mesh away from Γ suffices for the LBB condition to be satisfied. Adding wavelets to the trial spaces up to the level difference $L-1$ only near the boundary can be considered as a *stabilization.* This is actually a convenient procedure when working exclusively in wavelet coordinates as suggested by the recent studies of adaptive schemes in [CDD1].

It should be noted that, since the discretizations on $\square$ and Γ are completely independent, the quantitative choice of the coarsest mesh size is somewhat ambiguous. But any rescaling in a concrete case will be automatically reflected by the behavior of the trace matrices, a fact that should be kept in mind when interpreting the numerical results in Section 4.3. Roughly speaking, the condition of the trace matrix will deteriorate when the local dimension of the traces of the trial spaces on Γ becomes too small relative to the corresponding local dimensions of the Lagrange multiplier spaces. Clearly these local dimensions also depend on the order of the trial spaces which is in full agreement with the role of δ in (4.2.45).

Finally it should be mentioned that there are also techniques to completely *circumvent* the LBB condition that have been developed in the past years. For the Lagrange multiplier approach for the above problem with Dirichlet boundary conditions on $\Gamma = \partial\Omega$, stabilization techniques based on wavelets have been employed in [Be3]. There are many other problems where stabilization based on wavelets has been useful, see e.g. [CaM, BCT].

This idea can be used also in domain decomposition methods for elliptic boundary value problems. Here one decomposes the underlying domain into several subdomains with varying discretizations. For non–overlapping variants of these methods, compatibility conditions at the interface boundaries may restrict the applicability. To circumvent this, there are stabilizing strategies which either add bubble functions on the interfaces or discrete L_2 norms [BM] which, however, depend on the finest discretization and, therefore have to be prescribed beforehand. In [BeK], we add certain inner products for function spaces of 'negative smoothness' which can, by the Riesz basis property (I) from Section 2, be evaluated in terms of wavelets. Furthermore, the asymptotic optimality of the corresponding linear operator can be shown [BeK]. Here one needs the wavelet characterization of function spaces defined on the skeleton of the domain decomposition. Since the skeleton is not a Lipschitz domain, one cannot apply usual trace theorems. In this context, the concept of *d*-sets provides the corresponding characterizations [DkK].

4.3 Numerical Studies

Finally a numerical example is presented to illustrate the role and behavior of the above trace matrices in Section 4.2.4 and to quantify the theoretical estimates on condition numbers in Section 4.1.2.

Consider the problem

$$\begin{aligned} -\Delta y + y &= 1 \quad \text{in } \Omega, \\ y &= 0 \quad \text{on } \Gamma. \end{aligned} \tag{4.3.1}$$

Here the domain Ω is an open disc with radius R,

$$\Omega := \{\mathbf{x} \in I\!R^2 : (x_1 - 1/2)^2 + (x_2 - 1/2)^2 < R\} \tag{4.3.2}$$

which is embedded into the fictitious domain $\square = (0,1)^2$.

To my knowledge the only quantitative results concerning the LBB condition refer to polygonal boundaries, piecewise linear trial functions and piecewise constant Lagrange multipliers [GG]. Therefore, the circle as a boundary is chosen here deliberately so that a direct exploitation of the traces of trial functions is hardly conceivable. The discretization of (4.3.1) is briefly sketched only to an extent needed to relate to the above results.

The corresponding saddle point problem is set up in the framework of the Lagrange Multiplier—Fictitious Domain Approach (4.2.13). Since the multiresolution spaces on the boundary Γ can be defined via periodization, $\ell_0 = 0$ is chosen as the coarsest level on the boundary. In order to guarantee the norm equivalence to hold for $\|\cdot\|_{H^s(\Gamma)}$ for both $s = \pm 1/2$, we have taken on Γ piecewise linear functions as primal generators such that $d_\Gamma = 2$ in (3.2.55) and $\tilde{d}_\Gamma = 4$ for the dual multiresolution. On $\square$, tensor products of the biorthogonal wavelets generated by functions of order $d_\square = 2$ in (4.2.30) (piecewise linears) and order $\tilde{d}_\square = 4$ for the dual multiresolution constructed in [DKU2] have been employed. The lowest level on $\square$ is therefore $j_0 = 3$. The blocks $\mathbf{A}$ and $\mathbf{B}$ in (4.1.20) with respect to the wavelet bases $\Psi^{j,\square}$ and $\Psi^{\ell,\Gamma}$ have up to diagonal scaling the form

$$\begin{aligned} \mathbf{A}[j,j] &= (\nabla\Psi^{j,\square}, \nabla\Psi^{j,\square})_{L_2(\square)} + (\Psi^{j,\square}, \Psi^{j,\square})_{L_2(\square)}, \\ \mathbf{B}[\ell,j] &= (\Psi^{\ell,\Gamma}, B\Psi^{j,\square})_{L_2(\Gamma)}. \end{aligned}$$

Note that $\mathbf{B}[\ell,j]$ is a finite section of the matrix $(\Psi^\Gamma, B\Psi^\square)_{L_2(\Gamma)}$ from (4.2.52) corresponding to the spaces Q_ℓ, Y_j.

Remark 4.17 *As a consequence of the norm equivalences (4.1.17) and (4.1.18), the scaled versions*

$$\begin{pmatrix} \mathbf{D}_{-1,\square}[j,j] & \mathbf{0} \\ \mathbf{0} & \mathbf{D}_{1/2,\Gamma}[\ell,\ell] \end{pmatrix} \begin{pmatrix} \mathbf{A}[j,j] & \mathbf{B}[\ell,j]^T \\ \mathbf{B}[\ell,j] & \mathbf{0} \end{pmatrix} \begin{pmatrix} \mathbf{D}_{-1,\square}[j,j] & \mathbf{0} \\ \mathbf{0} & \mathbf{D}_{1/2,\Gamma}[\ell,\ell] \end{pmatrix}$$

$$= \begin{pmatrix} (\mathbf{D}_{-1,\square}\mathbf{A}\mathbf{D}_{-1,\square})[j,j] & \mathbf{B}_{1/2}[\ell,j]^T \\ \mathbf{B}_{1/2}[\ell,j] & \mathbf{0} \end{pmatrix}, \tag{4.3.3}$$

where again $\mathbf{B}_{1/2}[\ell,j]$ is a finite section of the matrix $\mathbf{B}_{1/2}$ defined in (4.2.55), establish uniformly boundedly invertible mappings on ℓ_2, provided that the LBB condition holds.

In practice, one would not compute the right hand side of (4.3.3) explicitly since the blocks are not quite as sparse as the usual nodal basis representations. Instead one computes first corresponding blocks with respect to the single scale bases $\Phi_j^\square$, Φ_ℓ^Γ. Recalling

(3.2.22), applying the blocks in (4.3.3) can then be achieved by applying the respective Fast Wavelet Transforms $\mathbf{T}^{\square}$ and $\mathbf{T}^{\Gamma}$ to the operator represented in terms of $\Phi_j^{\square}$, Φ_j^{Γ}, cf. (3.2.74). Furthermore, recall from Remark 3.1 that under the above locality assumptions on the wavelets, only $\mathcal{O}(\dim Y_j)$, respectively, $\mathcal{O}(\dim Q_\ell)$ arithmetic operations are needed for applying these transforms. In the present case, this means to compute

$$\mathbf{B}[\ell,j] = (\Psi^{\ell,\Gamma}, B\Psi^{j,\square})_{L_2(\Gamma)} = (\mathbf{T}_\ell^{\Gamma})^T\,(\Phi_\ell^{\Gamma}, B\Phi_j^{\square})_{L_2(\Gamma)}\,\mathbf{T}_j^{\square} \tag{4.3.4}$$

combined with a diagonal scaling according to (4.2.55) with $s = 1/2$. Recall that the Fast Wavelet Transforms $\mathbf{T}^{\square}$ and $\mathbf{T}^{\Gamma}$ contain the mask coefficients of the wavelets in terms of the single–scale functions, see Section 3.2.1 for their explicit form.

Since the circle

$$\Gamma = \{\mathbf{x} \in I\!R^2 :\ (x_1 - 1/2)^2 +\ (x_2 - 1/2)^2 = R\} \tag{4.3.5}$$

has the parametric representation

$$\boldsymbol{\kappa}(\tau) = \begin{pmatrix} x_1 = \kappa_1(\tau) := (1 + R\cos 2\pi\tau)/2 \\ x_2 = \kappa_2(\tau) := (1 + R\sin 2\pi\tau)/2 \end{pmatrix}, \quad \tau \in [0,1], \tag{4.3.6}$$

the computation of $\mathbf{B}_{1/2}$ therefore reduces to computing quantities of the form

$$(\phi_{\ell,k}^{\Gamma}, B\phi_{j,k'}^{\square})_{L_2(\Gamma)} = \frac{\pi}{2}\int_0^1 \phi_{j,k'}^{\square}(\boldsymbol{\kappa}(\tau))\,\phi_{\ell,k}^{\Gamma}(\tau)\,d\tau. \tag{4.3.7}$$

Approximating $\phi_{j,k'}^{\square}(\boldsymbol{\kappa}(\tau))$ by suitable linear combinations of the dual generators $\tilde{\phi}_{r,l}^{\Gamma}$ corresponding to the basis $\tilde{\Phi}^{\Gamma}$ for sufficiently high r, one can then use biorthogonality of the generator bases $\tilde{\Phi}_\ell^{\Gamma}$, Φ_ℓ^{Γ} to compute the resulting approximate integrals exactly [DM2, DKS1, K2]. For the computations reported below, the choice $r = 9$ has been sufficiently good to show no different quantitative behavior of the solution.

The above strategy has been used for the numerical tests which are discussed next.

Recall from Theorem 4.15 and (4.2.54) that the validity of the LBB condition can be inferred from the spectral properties of the (infinite) matrices $\mathbf{B}_{1/2}$ and $\mathbf{B}_{\delta+1/2}$. As pointed out above, the finite sections $\mathbf{B}_{1/2}[\ell,j]$ naturally appear in the wavelet representation (4.3.3) of the discrete problem (4.1.31). Moreover, finite sections $\mathbf{B}_{\delta+1/2}[\ell,j]$ are by (4.2.55) obtained by diagonally scaling $\mathbf{B}_{1/2}[\ell,j]$. It is therefore natural to exploit these circumstances for deriving quantitative information about the constants $c_{\delta+1/2,\Gamma} = \sigma_{\min}(\mathbf{B}_{\delta+1/2})$, $C_{1/2,\Gamma} = \sigma_{\max}(\mathbf{B}_{1/2})$. Ideally one would hope that the values $\sigma_{\min}(\mathbf{B}_{\delta+1/2}[\ell,j])$, $\sigma_{\max}(\mathbf{B}_{1/2}[\ell,j])$ provide good estimates for the quantities $c_{\delta+1/2,\Gamma}$, $C_{1/2,\Gamma}$ already for moderate values of ℓ and j in order to monitor the stability of the discretization, i.e., to get estimates for K_δ from (4.2.40) and hence L in (4.2.45). Some first experiments in this direction are recorded in Tables 4.1, 4.2 and 4.3. We present only the results for $R = 0.5$, i.e., for the case when Γ touches the boundary of $\square$. Analogous results for smaller R are dispensed with since moderately decreasing R shows essentially the same behavior. As expected, only when R becomes too small, which means that the coarsest mesh size on Γ becomes too small relative to the mesh size on $\square$, the smallest singular values $\sigma_{\min}(\mathbf{B}_{\delta+1/2}[\ell,j])$ deteriorate.

ℓ	N_Γ	j	$N_\square$	$\sigma_{\max}(\mathbf{B}_{\frac{1}{2}})$	$\sigma_{\max}(\mathbf{B}_{\frac{1}{2}+\delta})$	$\sigma_{\min}(\mathbf{B}_{\frac{1}{2}+\delta})$	$\kappa(\mathbf{B}_{\frac{1}{2}+\delta})$	$K_\delta[\ell,j]$	$L[\ell,j]$
3	8	5	1089	15.78	13.64	2.18	6.25	7.23	28.54
3	"	6	4225	22.81	19.37	3.10	6.24	7.35	28.78
3	"	7	16641	28.88	24.00	4.04	5.94	7.14	28.36
4	16	5	1089	23.29	21.59	2.29	9.45	10.19	33.50
4	"	6	4225	33.04	30.27	3.14	9.63	10.51	33.94
4	"	7	16641	40.16	36.28	4.19	8.66	9.59	32.61
5	32	5	1089	28.53	28.35	2.43	11.67	11.74	35.54
5	"	6	4225	46.74	46.08	3.23	14.25	14.46	38.54
5	"	7	16641	65.77	64.18	4.01	15.99	16.39	40.35
6	64	5	1089	42.24	44.81	2.53	17.74	16.72	40.63
6	"	6	4225	71.16	74.82	3.34	22.39	21.30	44.13
6	"	7	16641	90.57	94.81	4.13	22.97	21.95	44.56
7	128	5	1089	45.89	50.02	1.24	40.41	37.07	52.12
7	"	6	4225	99.54	109.76	3.36	32.63	29.59	48.87
7	"	7	16641	141.62	155.21	4.07	38.10	34.76	51.20

Table 4.1: Singular values and condition numbers of $\mathbf{B}_{1/2} = \mathbf{B}_{1/2}[\ell,j]$, $\mathbf{B}_{1/2+\delta} = \mathbf{B}_{1/2+\delta}[\ell,j]$, $K_\delta[\ell,j]$ for $\delta = 0.1$, and $L[\ell,j]$; $d_\square = 2$, $\tilde{d}_\square = 4$.

The experiments below involve refinement levels up to $\ell, j = 7$ on Γ and $\square$, respectively, see Tables 4.1, 4.2 and 4.3. To give an impression of the corresponding block sizes we also list the numbers N_Γ, $N_\square$ of the respective degrees of freedom. In order to economize on space we briefly write $\sigma_{\min/\max}(\mathbf{B}_s)$ instead of $\sigma_{\min/\max}(\mathbf{B}_s[j,\ell])$ in the table header. As mentioned before, the condition number $\kappa(\mathbf{B}_{\delta+1/2})$ should provide an indication for the behavior of K_δ. Therefore, we include the values $\kappa(\mathbf{B}_{\delta+1/2}[\ell,j])$ for comparison with the approximations

$$K_\delta[\ell,j] := \frac{\sigma_{\max}(\mathbf{B}_{1/2}[\ell,j])}{\sigma_{\min}(\mathbf{B}_{\delta+1/2}[\ell,j])}, \qquad L[\ell,j] := \frac{\log_2 K_\delta[\ell,j]}{\delta}.$$

Before discussing the results, a word on the choice of ℓ, j in Tables 4.1, 4.2 and 4.3 is in order. The issue is to get first sufficiently accurate information on the spectral properties of the infinite matrices $\mathbf{B}_s$ for $s = 1/2, 1/2+\delta$ through corresponding finite sections depending on ℓ, j. The resulting estimates $L[\ell,j]$ are then supposed to identify appropriate discretization levels for Γ and $\square$. To this end, note that we can hope to approximate $\sigma_{\min/\max}(\mathbf{B}_s) = \sqrt{\lambda_{\min/\max}(\mathbf{B}_s\mathbf{B}_s^T)}$ through finite sections of the matrices $\mathbf{B}_s\mathbf{B}_s^T$ which are given by

$$(\mathbf{B}_s[\ell,\infty])\,(\mathbf{B}_s[\ell,\infty]^T).$$

Thus, independently of the LBB condition, j should be large compared to ℓ in this context. Nevertheless, to see the effect of varying the 'inner' truncation level j, we

ℓ	N_Γ	j	$N_\square$	$\sigma_{\max}(\mathbf{B}_{\frac{1}{2}})$	$\sigma_{\max}(\mathbf{B}_{\frac{1}{2}+\delta})$	$\sigma_{\min}(\mathbf{B}_{\frac{1}{2}+\delta})$	$\kappa(\mathbf{B}_{\frac{1}{2}+\delta})$	$K_\delta[\ell,j]$	$L[\ell,j]$
3	8	5	1089	15.78	7.68	0.71	10.80	22.21	8.95
3	"	6	4225	22.81	10.31	0.99	10.36	22.93	9.04
3	"	7	16641	28.88	12.00	1.26	9.49	22.84	9.03
4	16	5	1089	23.29	16.05	0.74	21.70	41.76	8.97
4	"	6	4225	33.04	21.72	1.01	21.61	32.87	10.08
4	"	7	16641	40.16	25.01	1.31	19.05	30.59	9.87
5	32	5	1089	28.53	27.81	0.78	35.66	36.59	10.39
5	"	6	4225	46.74	44.21	1.00	44.11	46.64	11.09
5	"	7	16641	65.77	59.73	1.21	49.26	54.24	11.52
6	64	5	1089	42.24	57.42	0.80	71.33	52.48	11.43
6	"	6	4225	71.16	93.90	1.03	91.25	69.15	12.22
6	"	7	16641	90.57	118.03	1.22	96.63	74.15	12.42
7	128	5	1089	45.89	72.79	0.61	119.56	75.38	12.47
7	"	6	4225	99.54	168.04	1.04	161.88	95.88	13.17
7	"	7	16641	141.62	234.70	1.21	193.20	116.58	13.73

Table 4.2: Singular values and condition numbers of $\mathbf{B}_{1/2} = \mathbf{B}_{1/2}[\ell,j]$, $\mathbf{B}_{1/2+\delta} = \mathbf{B}_{1/2+\delta}[\ell,j]$, $K_\delta[\ell,j]$ for $\delta = 0.5$, and $L[\ell,j]$; $d_\square = 2$, $\tilde{d}_\square = 4$.

list the cases $j = 5,6,7$. Of course, one expects that the most reliable estimates are obtained when ℓ is small compared to j. In this case, the estimates $L[\ell,j]$ exhibit indeed the smallest values. However, the condition numbers $\kappa(\mathbf{B}_{\delta+1/2}[\ell,j])$ are in this regime smaller than $K_\delta[\ell,j]$. This reverses when ℓ and j are about the same.

Recall that the aim of testing different choices of j and ℓ is to obtain an impression of how accurate information about the spectral properties of the (infinite) trace matrices $\mathbf{B}_{1/2}, \mathbf{B}_{\delta+1/2}$ can be derived from low order discretizations like (4.3.3). At the same time we wish to see which of the family of estimates depending on δ provides the best information. Noting that for the present choice of Ψ^Γ, $\Psi^\square$ our theory covers $\delta \leq 1$, we record for $\delta = 0.1, 0.5$ and 1.0 the data in Tables 4.1, 4.2 and 4.3. The results clearly show that the smallest and largest singular values of the finite sections of the scaled trace matrices still vary significantly for all δ when changing the size of the sections in terms of j and ℓ. Thus, the numbers indicate that reliable information about the true value of K_δ seems to require higher values of ℓ and j with j significantly larger than ℓ. Nevertheless, larger δ is seen to produce for all choices of ℓ, j the smallest indicators $L[\ell,j]$ for sufficient differences of levels. Although they are not sufficiently good approximations for L, the tests confirm the expected trend that larger δ yields better criteria for the LBB condition. It would be interesting to choose even more regular dual systems and also trial spaces of order larger than two in order to increase δ. Since already the present simple cases confirm the trend, we have dispensed at this point with further more sophisticated implementations.

ℓ	N_Γ	j	$N_\Box$	$\sigma_{\max}(\mathbf{B}_{\frac{1}{2}})$	$\sigma_{\max}(\mathbf{B}_{\frac{1}{2}+\delta})$	$\sigma_{\min}(\mathbf{B}_{\frac{1}{2}+\delta})$	$\kappa(\mathbf{B}_{\frac{1}{2}+\delta})$	$K_\delta[\ell,j]$	$L[\ell,j]$
3	8	5	1089	15.78	3.78	0.18	21.60	90.12	6.49
3	"	6	4225	22.81	4.88	0.24	20.09	93.86	6.55
3	"	7	16641	28.88	5.44	0.30	17.93	95.25	6.57
4	16	5	1089	23.29	11.20	0.18	61.86	128.64	7.01
4	"	6	4225	33.04	14.79	0.24	60.50	135.21	7.08
4	"	7	16641	40.16	16.56	0.31	52.72	127.87	7.00
5	32	5	1089	28.53	27.44	0.19	144.74	150.49	7.23
5	"	6	4225	46.74	42.96	0.24	180.96	196.85	7.62
5	"	7	16641	65.77	56.77	0.28	201.58	233.54	7.87
6	64	5	1089	42.24	79.61	0.19	410.76	217.97	7.77
6	"	6	4225	71.16	120.75	0.32	374.03	220.44	8.65
6	"	7	16641	90.57	161.27	0.28	579.93	325.68	8.35
7	128	5	1089	45.89	129.88	0.13	969.25	342.48	8.42
7	"	6	4225	99.54	308.47	0.24	1266.84	408.78	8.68
7	"	7	16641	141.62	479.76	0.21	2305.99	680.68	8.56

Table 4.3: Singular values and condition numbers of $\mathbf{B}_{1/2} = \mathbf{B}_{1/2}[\ell, j]$, $\mathbf{B}_{1/2+\delta} = \mathbf{B}_{1/2+\delta}[\ell, j]$, $K_\delta[\ell, j]$ for $\delta = 1.0$, and $L[\ell, j]$; $d_\Box = 2$, $\tilde{d}_\Box = 4$.

The above choice of parameters ℓ, j so far provide indicators of questionable practical reliability. Thus, we proceed now discussing their influence on the condition numbers on the involved matrices appearing in (4.3.3) and the effect of preconditioning. We concentrate on the case when Ω is as large as possible relative to $\Box$, that is, when the radius is $R = 0.5$.

In Table 4.4, spectral condition numbers of the finite-dimensional operators appearing in the saddle point formulation (4.1.31) are displayed in order to test the effects of preconditioning and the validity of the LBB condition, depending on the resolution levels j and ℓ on $\Box$ and Γ, respectively. The condition numbers are determined as $\kappa(\mathbf{K}) = \lambda_{\max}(\mathbf{K})/\lambda_{\min}(\mathbf{K})$, where $\mathbf{K}$ is symmetric and positive definite. The maximal and minimal eigenvalues of $\mathbf{K}$ are computed by means of power and inverse power iterations.

The operator A represented in terms of the generator basis is

$$\mathbf{A}_\Phi = \mathbf{A}_\Phi[j, j] = (\nabla\Phi_j^\Box, \nabla\Phi_j^\Box)_{L_2(\Box)} + (\Phi_j^\Box, \Phi_j^\Box)_{L_2(\Box)}. \tag{4.3.8}$$

Its entries are computed exactly using refinement relations for the generators as described in [DM2, DKS1, K2]. Recall that $\mathbf{A}_\Phi$ can be asymptotically optimally preconditioned by applying the Fast Wavelet Transform (3.2.74) together with (3.2.75), compare the columns 4 and 5 in Table 4.4, or Table 3.2. As mentioned already in Section 3.2.4, one can even improve the constants in the condition number estimates by exchanging the

j	$N_\square$	ℓ	N_Γ	$\kappa(\mathbf{A}_\Phi)$	$\kappa(\mathbf{C}^{1/2}\mathbf{A}_\Phi\mathbf{C}^{1/2})$	$\kappa(\mathbf{B}_\Phi\mathbf{B}_\Phi^T)$	$\kappa(\mathbf{S}_\Phi)$	$\kappa(\mathbf{S}_\Psi)$
3	81	3	8	$3.12e+02$	5.79	$4.81e+00$	$1.52e+02$	$9.84e+01$
3	"	4	16	"	"	$9.63e+00$	$5.20e+02$	$1.56e+02$
3	"	5	32	"	"	$1.33e+02$	$9.11e+03$	$7.00e+02$
3	"	6	64	"	"	$6.81e+16$	$6.53e+17$	$2.72e+16$
4	289	3	8	$1.14e+03$	6.75	$4.04e+00$	$1.35e+02$	$9.00e+01$
4	"	4	16	"	"	$4.89e+00$	$3.44e+02$	$1.16e+02$
4	"	5	32	"	"	$9.52e+00$	$1.31e+03$	$1.59e+02$
4	"	6	64	"	"	$2.87e+02$	$4.02e+04$	$1.52e+03$
4	"	7	128	"	"	$1.02e+17$	$8.28e+17$	$2.88e+16$
5	1089	3	8	$4.34e+03$	7.54	$3.77e+00$	$1.29e+02$	$8.69e+01$
5	"	4	16	"	"	$3.82e+00$	$2.97e+02$	$1.07e+02$
5	"	5	32	"	"	$4.85e+00$	$7.30e+02$	$1.22e+02$
5	"	6	64	"	"	$7.94e+00$	$2.70e+03$	$1.36e+02$
5	"	7	128	"	"	$1.23e+03$	$4.07e+05$	$7.07e+03$
5	"	8	256	"	"	$3.50e+17$	$2.85e+18$	∞
6	4225	3	8	$1.69e+04$	7.86	$3.68e+00$	$6.21e+04$	x
6	"	4	16	"	"	$3.60e+00$	$6.01e+04$	x
6	"	5	32	"	"	$3.78e+00$	$6.38e+04$	x
6	"	6	64	"	"	$4.61e+00$	$7.79e+04$	x
6	"	7	128	"	"	$8.66e+00$	$1.46e+05$	x
6	"	8	256	"	"	$4.38e+03$	$7.40e+07$	x
6	"	9	512	"	"	∞	∞	x
7	16641	3	8	$6.66e+04$	8.39	$3.64e+00$	$2.42e+05$	x
7	"	4	16	"	"	$3.50e+00$	$2.33e+05$	x
7	"	5	32	"	"	$3.31e+00$	$2.20e+05$	x
7	"	6	64	"	"	$3.63e+00$	$2.41e+05$	x
7	"	7	128	"	"	$4.23e+00$	$2.82e+05$	x
7	"	8	256	"	"	$7.38e+00$	$4.92e+05$	x
7	"	9	512	"	"	$7.48e+03$	$4.98e+08$	x
8	66049	3	8	$2.64e+05$	8.82	$3.63e+00$	$9.58e+05$	x
8	"	4	16	"	"	$3.45e+00$	$9.11e+05$	x
8	"	5	32	"	"	$3.19e+00$	$8.42e+05$	x
8	"	6	64	"	"	$3.19e+00$	$8.42e+05$	x
8	"	7	128	"	"	$3.58e+00$	$9.45e+05$	x
8	"	8	256	"	"	$3.92e+00$	$1.03e+06$	x
8	"	9	512	"	"	$4.87e+00$	$1.29e+06$	x

Table 4.4: Spectral condition numbers of operators in the saddle point problem (4.2.13).

diagonal scaling matrices by the diagonal of $\mathbf{A}_{\Psi^J}$. An alternative is to use the BPX–preconditioner [DK1, O1], see [K1] for a detailed description. The boundary operator is also first set up with respect to the generator bases $\Phi_j^{\square}, \Phi_\ell^{\Gamma}$ as in (4.3.4),

$$\mathbf{B}_\Phi = \mathbf{B}_\Phi[\ell, j] = (\Phi_\ell^\Gamma, B\Phi_j^\square)_{L_2(\Gamma)}. \tag{4.3.9}$$

The condition number of $\mathbf{B}_\Phi \mathbf{B}_\Phi^T$ provides information on the singular values of $\mathbf{B}_\Phi$ and therefore on the constant in the LBB condition (4.2.26) when ℓ grows relative to j. In order to get an idea on the performance of the Uzawa algorithm (4.1.32), we have also computed the condition numbers of the Schur complement

$$\mathbf{S}_\Phi = \mathbf{S}_\Phi[\ell, j] = \mathbf{B}_\Phi[\ell, j]\ (\mathbf{A}_\Phi[j, j])^{-1}\ (\mathbf{B}_\Phi[\ell, j])^T. \tag{4.3.10}$$

For $j > 5$, the term $\kappa(\mathbf{S}_\Phi)$ has been estimated as

$$\kappa(\mathbf{S}_\Phi) \leq \kappa(\mathbf{A}_\Phi[j, j])\, \kappa\left(\mathbf{B}_\Phi[\ell, j]\, (\mathbf{B}_\Phi[\ell, j])^T\right), \tag{4.3.11}$$

see e.g. [Br], Chapter 4.5. In the last column, the condition number of the Schur complement which is preconditioned by applying the Fast Wavelet Transform $\mathbf{T}_\ell^\Gamma$ on Γ (3.2.23) including diagonal scaling,

$$\mathbf{S}_\Psi = \mathbf{S}_\Psi[\ell, j] = (\mathbf{T}_\ell^\Gamma)^T\, (\mathbf{S}_\Phi[\ell, j])\, \mathbf{T}_\ell^\Gamma, \tag{4.3.12}$$

is displayed. This is done by explicitly setting up the Schur complement in terms of the wavelet basis. Since this was not possible for higher levels because of storage restrictions, the sign 'x' in Table 4.4 means that the values have not been computed. Furthermore, the term '∞' indicates that the power iterations did not converge after several thousand iterations. The case $j \geq 3$, $\ell = 7$ is not admissible for this example since the number of rows of B is bigger than the number of columns.

The remainder of the setup of the operators is done as before.

The results displayed in Table 4.4 can be interpreted as follows. As expected, the condition number of the stiffness matrix with respect to the single–scale basis, $\mathbf{A}_\Phi$, grows like 2^{2j} as j increases (fifth column) which reduces to a small constant after preconditioning (sixth column). It can be seen in the seventh column that the condition number of $\mathbf{B}_\Phi \mathbf{B}_\Phi^T$ is moderately small as long as $\ell \leq j$. Even the case $\ell = j+1$ which violates the sufficient conditions derived in Theorem 4.15 for the validity of the LBB condition gives still small acceptable constants in all cases. Only when $\ell = j+2$, the condition number starts to be in the same order of magnitude as the unpreconditioned $\mathbf{A}_\Phi$, indicating that the violation of the LBB condition will have an effect on the numerical computations. The results become unacceptable for computations when ℓ is too large relative to j, i.e., the cases when $\ell \geq j+3$. The eighth column displays the condition numbers of the unpreconditioned Schur complement $\mathbf{S}_\Phi$. In view of (4.3.11) the growth of the condition number of $\mathbf{S}_\Phi$ seems to be affected more by the condition number of $\mathbf{A}_\Phi$ than by the violation of the LBB condition quantified by $\mathbf{B}_\Phi \mathbf{B}_\Phi^T$. Preconditioning the Schur complement by employing the Fast Wavelet Transform including diagonal scaling provides an improvement in orders of magnitude in the computed cases, where the results get better for higher levels j, ℓ. As observed before, taking instead of the diagonal scaling the inverse of the diagonal of $\mathbf{S}_\Psi$ would further improve the absolute values.

To validate whether the above numerical indicators are realistic or too pessimistic, we discuss now the behavior of an iterative solution of (4.1.31) for different choices of ℓ and j. Since by Remark 4.17 the Schur complement of the right hand side of (4.3.3) has uniformly bounded condition numbers when the LBB condition holds, we have employed a CG–Uzawa algorithm. By this we mean that Uzawa's method (4.1.32) (see also (6.6.50) below for more details) is used as outer iteration for solving the system (4.1.31) with the CG–method for approximating the inverse of $(\mathbf{D}_{-1,\Box}\mathbf{A}\mathbf{D}_{-1,\Box})[j,j]$ up to a certain tolerance in an inner iteration. The results are displayed in Table 4.5. The stopping criterion is based on the ℓ_2 norm of (the dual wavelet coefficients of) the residual

$$\mathbf{r}_\Lambda := \begin{pmatrix} \mathbf{A}_\Lambda \mathbf{y}_\Lambda + \mathbf{B}_\Lambda^T \mathbf{p}_\Lambda - \mathbf{f}_\Lambda \\ \mathbf{B}_\Lambda \mathbf{y}_\Lambda - \mathbf{u}_\Lambda \end{pmatrix}$$

which by (4.1.22) is proportional to the error of (y_Λ, p_Λ) in $H^1(\Box)$ and $H^{-1/2}(\Gamma)$. Since we use a second order discretization for both quantities, the first numbers in the third column reveal the total number of CG iterations necessary to force the ℓ_2 error of the residual to be smaller than `tol`$= \min\{2^{-j}, 2^{-\ell}\}$. The inner iterations are terminated when the error is smaller than $0.01*$`tol`. The numbers in parentheses show the number of Uzawa iterations. The ℓ_2 error of the residual is displayed in the fourth column. In the next column `sec` is the amount of CPU time computed on a Silicon Graphics work station with a Silicon Graphics C^{++} compiler. The last column shows the ℓ_2 error of the approximate solution $\mathbf{y}_\Lambda$ restricted to the boundary Γ. This serves as an indicator how well the Lagrange–Multiplier approach enforces the homogeneous Dirichlet boundary conditions in the strong sense. This error is computed by sampling the expanded approximated solution at the boundary approximated relative to level 10. In comparison to samples on even finer discretizations, this has turned out to be accurate enough.

Assessing the amount of iterations displayed in Table 4.5 yields that the preconditioned Uzawa scheme is relatively stable as long as $\ell \leq j+1$. Only when the level ℓ on Γ exceeds the level j on $\Box$ by more than 1 in this example, the significant increase of necessary iterations indicates a severe violation of the LBB condition. This tendency confirms the previous interpretations and demonstrates that the above indicators in terms of the trace matrices are indeed too pessimistic. I would like to point out that there are several places in the setup, construction of the wavelets and implementation which could be improved so that the absolute iteration numbers could further be reduced.

A number of comments on the accuracy of the approximation that can be drawn from Table 4.5 are in order. The tolerance `tol`$= \min\{2^{-j}, 2^{-\ell}\}$ is chosen here as an attempt to balance the accuracy on the domain and the boundary. In contrast to the sufficient conditions for the validity of the LBB condition, the boundary conditions should be approximated on a somewhat finer grid. On the other hand, the discretization on the domain must allow for a better approximation. Thus, choosing the tolerance as the minimum of 2^{-j} and $2^{-\ell}$ rather than the maximum yields a smaller error for both the residual and the boundary approximation, as can be seen for the numbers for $j = 3$. In this case the tolerance is dominated by the increasing level ℓ on Γ, and the boundary approximation displayed in the last column is roughly halved as expected by the second order approximation. Of course, further increasing the level on Γ would eventually result in saturation because of the relatively coarse level on $\Box$. Increasing the level on $\Box$ yields

j $\square$	ℓ Γ	#it	$\|\mathbf{r}_\Lambda\|_{\ell_2}$	sec	$\|\mathbf{y}_\Lambda\|_\Gamma\|_{\ell_2}$
3	3	50(5)	0.0898359951	0.028	$1.3942832582e-04$
3	4	63(4)	0.0245526291	0.034	$3.9440941778e-05$
3	5	64(4)	0.0194782569	0.035	$2.7920984717e-05$
3	6	100(7)	0.0124709705	0.055	$1.3762230847e-05$
4	3	77(4)	0.0221207827	0.140	$1.0109135445e-04$
4	4	64(3)	0.0621973224	0.118	$7.2377426138e-05$
4	5	80(4)	0.0162813577	0.148	$2.9872219698e-05$
4	6	313(21)	0.0129720238	0.578	$1.4797850063e-05$
4	7	412(28)	0.0065149366	0.792	$6.3157077631e-06$
5	3	90(4)	0.0212095842	0.645	$1.1045265099e-04$
5	4	91(4)	0.0223583836	0.698	$5.2610190468e-05$
5	5	90(4)	0.0179197401	0.671	$2.9848911895e-05$
5	6	94(4)	0.0145541630	0.704	$1.8641440445e-05$
5	7	470(29)	0.0068858511	3.568	$7.4215577399e-06$
6	3	119(5)	0.0105475171	4.535	$1.1435482527e-04$
6	4	119(5)	0.0150313420	4.568	$5.3957049625e-05$
6	5	131(6)	0.0107629856	5.044	$2.1450571117e-05$
6	6	102(4)	0.0146560671	3.992	$1.8358210750e-05$
6	7	119(5)	0.0052573891	4.612	$7.7080275609e-06$

Table 4.5: Iteration numbers for solving (4.1.31) by (preconditioned) CG–Uzawa method; #it: total number of PCG iterations with number of Uzawa steps in parentheses until $\|\mathbf{r}_\Lambda\|_{\ell_2} \leq$`tol`; **sec**: CPU time in seconds; $\|\mathbf{y}_\Lambda|_\Gamma\|_{\ell_2}$: error of $\mathbf{y}_\Lambda$ restricted to Γ in ℓ_2.

a decreasing residual where, however, the effect of the different quantities is blurred. For level $j = 6$ when the tolerance is determined by 2^{-j} up to $\ell = 6$, one can detect the effect of reducing the mesh size on the boundary: the error for the approximation on the boundary is halved for increasing ℓ while the overall residual stays in the same range.

In summary, the Fictitious Domain–Lagrange Multiplier approach provides a surprisingly good approximation of $\mathbf{y}_\Lambda|_\Gamma$. In fact, the error $\|\mathbf{y}_\Lambda|_\Gamma\|_{\ell_2}$ seems to behave like $\mathcal{O}(\|\mathbf{r}_\Lambda\|_{\ell_2}^2)$.

The numbers are finally complemented by the following figures, Figures 4.1, 4.2 and 4.3, which give an impression of the shape of the solution of (4.3.1) in weak form (4.2.13). The solution is plotted as a surface plot over the domain $\square$. The resolution for all plots is level 7; higher resolutions could not be distinguished visually from these. Figures 4.1, 4.2 and 4.3 display plots of the solution computed on levels $j = 4, 5$ and $j = 6$, respectively. Each figure distinguishes further between approximation levels $\ell = 4$ (top) and $\ell = 6$ (bottom) on the boundary Γ. The approximations in Figure 4.1 are too coarse, although the higher approximation level $\ell = 6$ on Γ appears to be slightly better than level $\ell = 4$. The case $j = 5$ displayed in Figure 4.2 provides already a much better approximation of the boundary conditions on Γ so that the difference in levels $\ell = 4$ and $\ell = 6$ can hardly be distinguished. This can also be seen from the level curves. The plots displayed in 4.3 for $j = 6$ look sufficiently good and show no distinction between the different levels of resolution on Γ. For higher levels on $\square$, one cannot distinguish visually any further difference. Thus, for this example, choosing $j = 6$ and $\ell = 4$ seems to provide a sufficiently fine plot for approximating the solution of (4.3.1) by the Fictitious Domain—Lagrange Multiplier Approach.

In summary, one sees that refining the mesh for the Lagrange multipliers improves the accuracy of the boundary conditions. Of course, this effect is limited by the capability of the traces of the trial spaces on $\square$ to match the boundary conditions. Moreover, in order for the boundary conditions to be satisfied well the grid for the Lagrange multiplier has to be chosen sufficiently small. Recall that by the very nature of the trace matrices prescribing any level ℓ on Γ stability is already ensured when the trial spaces on $\square$ contain *only* those wavelets up to a corresponding somewhat higher scale $j = \ell + L - 1$ which overlap the boundary regardless of the mesh size in the interior of Ω away from the boundary. This corresponds to a graded mesh which is successively refined towards the boundary. In that sense the characterization of stability in terms of the trace matrices may open some promising perspectives for further investigations. In particular, it naturally blends into the analysis of adaptive concepts from [CDD1].

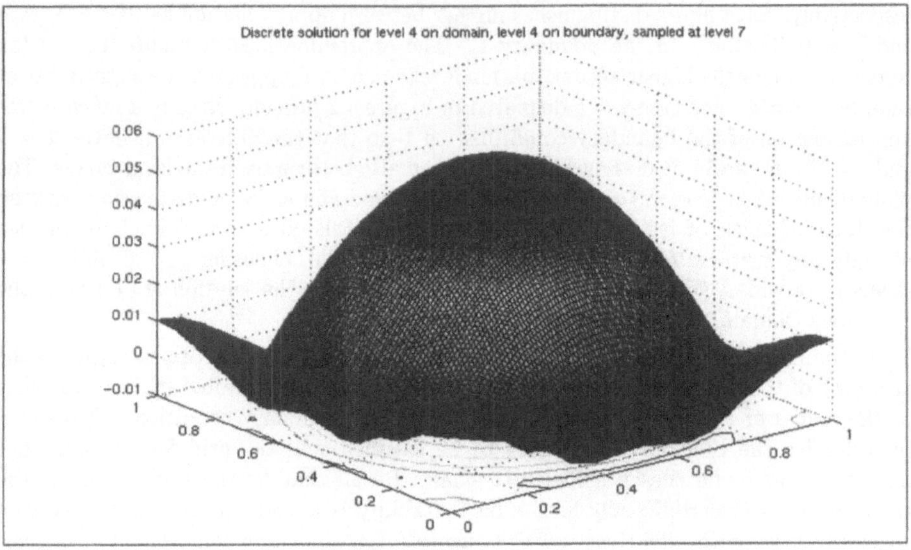

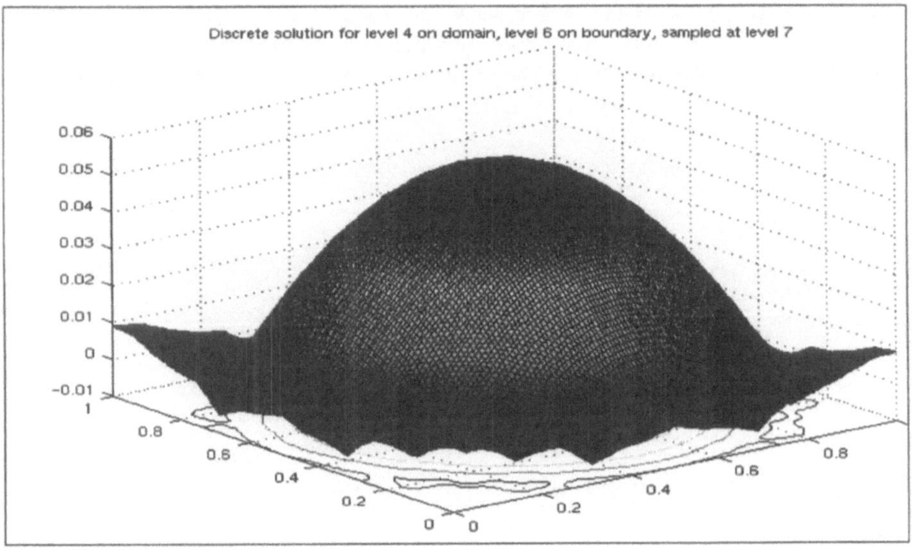

Figure 4.1: Solution of (4.3.1) in weak form (4.2.13) on refinement level $j = 4$ on $\Box$ and level $\ell = 4, 6$ on Γ, sampled on level 7.

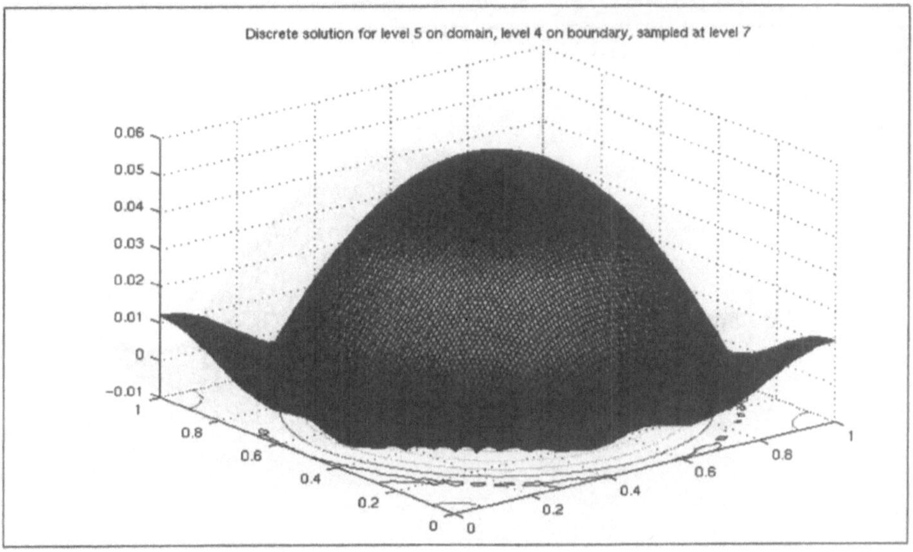

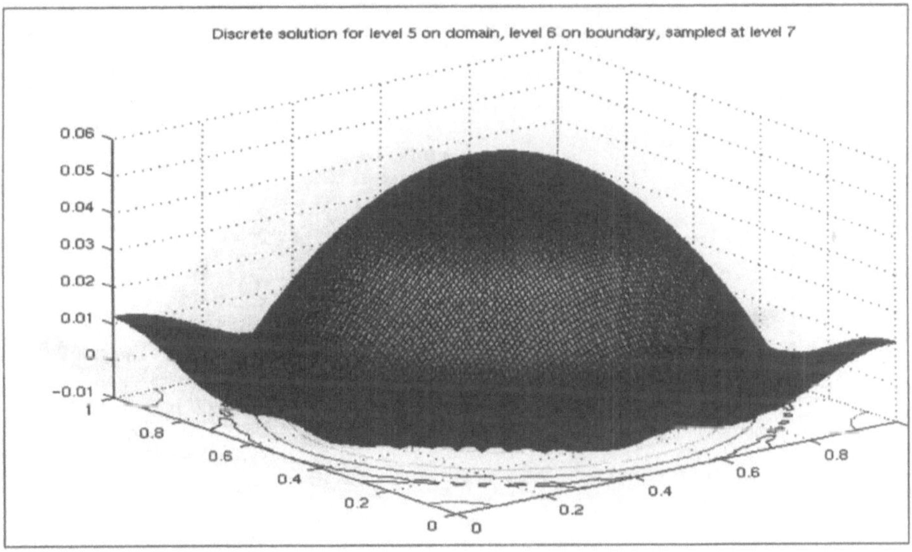

Figure 4.2: Solution of (4.3.1) in weak form (4.2.13) on refinement level $j = 5$ on $\square$ and level $\ell = 4, 6$ on Γ, sampled on level 7.

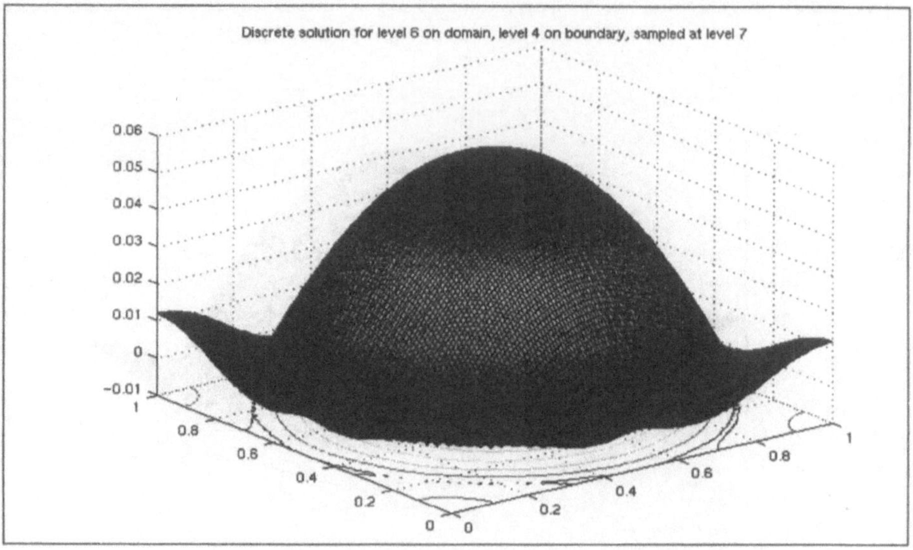

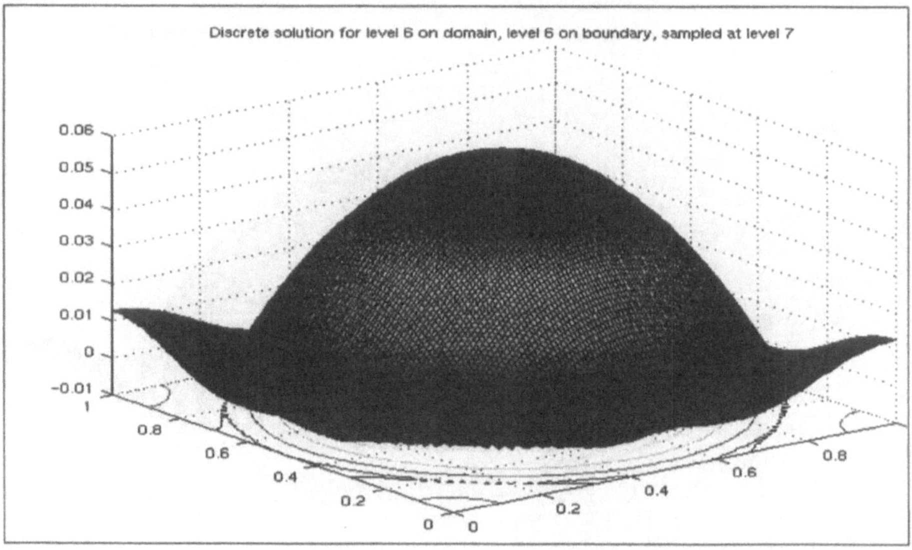

Figure 4.3: Solution of (4.3.1) in weak form (4.2.13) on refinement level $j = 6$ on $\Box$ and level $\ell = 4, 6$ on Γ, sampled on level 7.

5 Least Squares Problems

5.1 Introduction

In the past decade, the developments of efficient multilevel preconditioning techniques have revived the interest in least squares discretizations of boundary value problems, see e.g. [BG, BLP1, Sta]. They appear to be an interesting direction of research, although not without several opposing perspectives. On one hand, they are particularly tempting because, in principle, they allow to turn a variety of problems into a positive definite symmetric variational formulation. This may impose less stringent compatibility constraints on the underlying discretizations and also allows for the employment of the many available iterative methods for symmetric positive definite problems. Since, on the other hand, the approach inherently squares the problem, this may enhance ill-conditioning. This in turn might be avoided by suitable preconditioning techniques combined with a judicious choice of the least squares functional. However, these least squares functionals often turn out to be difficult to deal with in practice since they may require to evaluate norms for noninteger or even negative function spaces. Of course, the employment of new tools and concepts may well enhance the one or other advantage.

Therefore, the main objective of the recent study [DKS2] has been a possibly systematic treatment of the following three issues, all in the light of recent developments of wavelet concepts, namely:

(i) the identification of appropriate least squares functionals,

(ii) the numerical evaluation of resulting inner products,

(iii) the condition of corresponding discrete problems.

Some more comments on these issues are in order.

Ad (i): The importance of this point is already reflected by the number of investigations of this problem, for instance, in connection with the treatment of first order systems arising from second order elliptic boundary value problems, see e.g. [BLP1, BLP2]. In this context, the main guide line has been to avoid too stringent regularity requirements on the solution imposed by an improper choice of the least squares functional. These studies have been a major motivation for the recent investigations in [DKS2]. On one hand, the aim was to bring out the main governing mechanisms in this context and, on the other hand, we wanted to stress the potential of covering in fact a much wider scope of problems based on a single common principle. One focus is the application of least squares techniques to general saddle point problems of the form (4.1.9).

To this end, we consider in Section 5.2 like in (2.11) a general system of weakly defined operator equations $\mathcal{L}U = F$ which is assumed to establish an isomorphism $\|\mathcal{L}V\|_{\mathcal{H}'} \sim \|V\|_{\mathcal{H}}$. The point is that this mapping property always identifies a suitable least squares functional in terms of the involved dual norms in $\|\cdot\|_{\mathcal{H}'}$ or any equivalent quantities. This starting point for least squares methods fits exactly into the framework of the general setting from Chapter 2, namely, it is just Step 1 there.

Of course, relations like (2.13) (possibly involving other pairs of spaces than duality pairs) are fundamental for stability and have been identified in many individual cases. In most of those cases, the main effort is spent on how to deal with the dual norms appearing there. Here we stress that, once a duality isomorphism of the form (2.13) has been established, essentially the same machinery concerning (ii) described below applies to a variety of settings. Thus, only the natural regularity properties of the solution enter. Moreover, stability is obtained in a natural way and the discrete solutions exhibit optimal convergence rates.

In Section 5.2 the general format of the least squares formulations is described. Then in Section 5.3 as an example the general saddle point problems introduced in Section 4.1 are placed into this framework. This formulation includes, in particular, the Lagrange Multiplier—Fictitious Domain Approach (4.2.13). In [DKS2], different other examples have been recalled and the relevant terms identified, among them systems of first order equations, a transmission problem involving singular boundary integral operators, and (as a vector-valued example) a weak formulation of the Stokes problem with appended divergence *and* boundary conditions, thus, yielding a three-by-three block saddle point problem. In all these cases, it is verified that essentially the same mechanisms apply to ensure stable discrete problems. In fact, a common theme in all examples is to treat inhomogeneous (Dirichlet) boundary conditions 'separately' in the least squares formulation.

The fact that in the least squares formulation constraints like the LBB condition do *not* arise provides an interesting alternative to the investigations in Chapters 4.1.3 and 4.2.4. This is even the more so since the formulation of general saddle problems as positive definite symmetric problems opens promising perspectives to the employment of the adaptive strategies with optimal convergence rates from [CDD1].

Ad (ii): In many examples that can be found in the literature, one encounters dual norms and corresponding inner products that are hard to deal with practically, e.g., with noninteger or negative Sobolev indices. For instance, the complications introduced by the occurrence of the H^{-1} norm are well-known [BLP1, BLP2]. At this point, the recent advances in the developments of wavelet methods offer a common platform for dealing with difficult norms as well as with nonlocal operators such as singular integral operators. As a counterpart to (2.13), recall from Chapter 2 that appropriate wavelet bases induce isomorphisms between certain function spaces and weighted sequence spaces. This can be used to define discrete norms that are equivalent to the relevant norms arising in the least squares functionals like (2.18) or (2.22). Such wavelet expansions offer concrete Riesz maps in this context. The various ramifications of such facts have been discussed in Section 3. This suggests a common strategy for defining computable least squares functionals in all the cases under consideration.

The fact that dual norms such as the H^{-1} norm are equivalent to weighted sequence norms for wavelet expansion coefficients has been exploited earlier, for instance, in [Be1, DDHS] in connection with a-posteriori error estimators or for stabilizing semi-definite problems [Be2, Be3]. However, note that the equivalent norms provided by (2.18) still involve *infinite sums.* An important point addressed in Section 5.5 is therefore to *truncate* these sums in such a way that stability is preserved. Whenever the Galerkin discretization for the original system $\mathcal{L}U = F$ is already stable, there is no problem because it suffices to retain those summands in the ℓ_2 norms which correspond to basis functions in the trial spaces. This is the case for the transmission problem discussed in [DKS2]. We refer to this as *symmetric* truncation. For all other cases, it will be shown in Section 5.5 that appropriate *expanded* truncation ensures stability. This means that the discrete norms must involve certain more terms than those appearing in the trial spaces. However, the cardinality of the extended sets of indices still remains proportional to the dimension of the trial spaces, see also [Be2, Be3] for the case of uniform refinements in connection with stabilization techniques for semi-definite problems.

Ad (iii): Section 5.6 is devoted to a brief discussion of the resulting discrete problems. The upshot is that the combination of the two isomorphisms (2.13) and (2.18) always leads to positive definite quasi-sparse linear systems with uniformly bounded condition numbers, compare (3.2.75).

It has also been shown in [DKS2] that with a suitable truncation strategy one always retrieves optimal error estimates in the *energy norm*. In addition, we have revisited there the question of L_2 error estimates for first order elliptic systems. The main point is that in contrast to [BLP1] additional strong requirements on the preconditioner do *not* arise in the present context.

We conclude this chapter with a numerical example concerning the Fictitious Domain–Lagrange Multiplier Approach (4.2.13). Since this involves the H^{-1} norm as well as broken trace norms, it serves as a good illustration of the concept.

5.2 General Setting

The conceptual framework for the remainder of this chapter adheres to the notation introduced in Chapter 2 which will be briefly recalled next.

The general format of the operator equation (2.10) is the following: For given $F = (f_1, \dots, f_M)^T \in \mathcal{H}'$, find $U = (u_1, \dots, u_M)^T \in \mathcal{H}$ such that

$$\sum_{j=1}^{M} \mathcal{L}_{i,j} u_j = \mathcal{L}_i U = f_i, \quad i = 1, \dots, M. \tag{5.2.1}$$

This has been abbreviated in (2.11) as

$$\mathcal{L}U = F, \tag{5.2.2}$$

where $\mathcal{L} = (\mathcal{L}_{i,j})_{i,j=1}^{M}$. Recall also from Chapter 2 that $\mathcal{L}$ is well-posed if $\mathcal{L}$ is an isomorphism from $\mathcal{H}$ to $\mathcal{H}'$, i.e., there exist constants $0 < c_{\mathcal{L}} \le C_{\mathcal{L}} < \infty$ such that

$$c_{\mathcal{L}} \|V\|_{\mathcal{H}} \le \|\mathcal{L}V\|_{\mathcal{H}'} \le C_{\mathcal{L}} \|V\|_{\mathcal{H}}, \quad V \in \mathcal{H}, \tag{5.2.3}$$

holds. Sometimes this is also written as

$$\sum_{i=1}^{M} \|v_i\|_{H_i}^2 \sim \sum_{i=1}^{M} \|\mathcal{L}_i V\|_{H'_{i,0}}^2, \quad V \in \mathcal{H}. \tag{5.2.4}$$

We have considered in [DKS2] several examples where in contrast to systems of Agmon-Douglis-Nirenberg type [ADN] $\mathcal{L}$ involves *differential* as well as *integral* operators.

The relevance of (5.2.3) is that it determines a *natural least squares formulation* for the solution of (5.2.1), namely, to find U that

$$\text{minimizes} \quad \|\mathcal{L}V - F\|^2_{\mathcal{H}'} \quad \text{over} \quad \mathcal{H}.$$

On the other hand, these norms and the underlying inner products $(\cdot,\cdot)_{H'_{i,0}}$ may be numerically hard to evaluate. Therefore, we will identify suitable *equivalent norms*

$$\|\cdot\|_i^2 := (\cdot,\cdot)_i \quad \sim \quad \|\cdot\|^2_{H'_{i,0}}, \quad i = 1,\ldots,M, \tag{5.2.5}$$

which *can* be efficiently evaluated or appropriately approximated in a sense to be detailed later.

On account of (5.2.5), the solution U of (5.2.1) is the unique minimizer of the quadratic functional

$$LS(V) := \sum_{i=1}^{M} \|\mathcal{L}_i V - f_i\|_i^2 \longrightarrow \min. \tag{5.2.6}$$

This is apparently equivalent to solving the minimization problem

$$\operatorname{argmin}\{q(V) : V \in \mathcal{H}\}, \qquad q(V) := \tfrac{1}{2} O(V,V) - F(V), \tag{5.2.7}$$

with

$$\begin{aligned} O(V,W) &:= \sum_{i=1}^{M} (\mathcal{L}_i V, \mathcal{L}_i W)_i, \\ F(V) &:= \sum_{i=1}^{M} (\mathcal{L}_i V, f_i)_i. \end{aligned} \tag{5.2.8}$$

Remark 5.1 *By standard variational arguments, one knows that U solves (5.2.6) if and only if U is the solution of*

$$O(U,V) = F(V), \quad V \in \mathcal{H}. \tag{5.2.9}$$

Moreover, one concludes from (5.2.5) and (5.2.4) the stability estimates

$$\sum_{i=1}^{M} \|v_i\|^2_{H_i} \sim O(V,V), \quad V = (v_1,\ldots,v_M)^T \in \mathcal{H}, \tag{5.2.10}$$

i.e., the symmetric bilinear form $O(\cdot,\cdot)$ is $\mathcal{H}$-elliptic.

It will be instructive to identify the operator induced by $O(\cdot,\cdot)$. To this end, let the Riesz map $\mathcal{R}_i : H'_{i,0} \to H_{i,0}$ be defined by

$$\langle \mathcal{R}_i v, w\rangle = (v,w)_i, \quad v,w \in H'_{i,0}. \tag{5.2.11}$$

Roughly speaking, the Riesz map $\mathcal{R}_i$ undoes the shift in Sobolev scale caused by $\mathcal{L}_i$ and is, thus, a special case of the operator introduced in (3.2.62).

Then the solution U of (5.2.1) can be viewed as the solution of the operator equation

$$\mathcal{M}U = G, \tag{5.2.12}$$

where

$$\mathcal{M} := \sum_{i=1}^{M} \mathcal{L}_i' \mathcal{R}_i \mathcal{L}_i, \quad G =: \sum_{i=1}^{M} \mathcal{L}_i' \mathcal{R}_i f_i \tag{5.2.13}$$

and $\mathcal{M}$ is now positive definite and self-adjoint.

It will be described next how general saddle point problems of the type considered in Section 4.1.1 are written in this form. In fact, they can be transformed into an equivalent variational problem of the type (5.2.9) by introducing auxiliary inner products $(\cdot,\cdot)_i$ based on wavelet expansions which satisfy (5.2.5). Since these quantities still involve infinite sums we will then show later in Section 5.5 that appropriate truncation preserves the stability (5.2.10).

5.3 Least Squares Formulation of General Saddle Point Problems

We see now that the general saddle point problems described in Section 4.1.1 fit naturally into the framework of Section 5.2. We identify the relevant spaces and operators to place them into the notation of Section 5.2. The Hilbert spaces H_1, H_2 are

$$H_1 = Y, \qquad H_2 = Q$$

with corresponding norms and inner products, and continuous bilinear forms $a(\cdot,\cdot)$, $b(\cdot,\cdot)$ defined on $H_1 \times H_1$, $H_1 \times H_2$, respectively. The operators A and B, B' are defined as in (4.1.3) and (4.1.5). Furthermore, let

$$H_{2,0} := H_2 / \ker B'. \tag{5.3.1}$$

Recall that the minimization of a quadratic functional of the type $\frac{1}{2}a(v,v) - \langle f, v\rangle$ over some subspace $H_{1,0} \subseteq H_1$ under the constraint $b(v,q) = \langle g, q\rangle$ leads to the following saddle point problem: given $(f,u) \in H_{1,0}' \times H_{2,0}'$, find $(y,p) \in H_{1,0} \times H_{2,0}$ such that

$$\begin{aligned} a(y,v) + b(v,p) &= \langle f, v\rangle \qquad \text{for all } v \in H_{1,0},\\ b(y,q) &= \langle u, q\rangle \qquad \text{for all } q \in H_{2,0}, \end{aligned} \tag{5.3.2}$$

holds, cf. (4.2.10). In operator form, (5.3.2) is rewritten as

$$\begin{pmatrix} A & B' \\ B & 0 \end{pmatrix} \begin{pmatrix} y \\ p \end{pmatrix} = \begin{pmatrix} f \\ u \end{pmatrix} \tag{5.3.3}$$

which is abbreviated as

$$\mathcal{L}U = F. \tag{5.3.4}$$

We then have the following result.

Corollary 5.2 *Under the assumptions of Theorem 4.1, the operator*

$$\mathcal{L} = \begin{pmatrix} A & B' \\ B & 0 \end{pmatrix} \; : \; \mathcal{H} \to \mathcal{H}' \tag{5.3.5}$$

given by (5.3.3) satisfies (5.2.3) for

$$\mathcal{H} = H_{1,0} \times H_{2,0}.$$

Moreover, the problem (5.3.2) is of the form (5.2.1), (5.2.4) for $M = 2$, and a corresponding least squares functional

$$LS(v,q) := \|Av + B'q - f\|_1^2 + \|Bv - u\|_2^2 \tag{5.3.6}$$

is on account of (5.2.5) equivalent to

$$\|Av + B'q - f\|_{H'_{1,0}}^2 + \|Bv - u\|_{H'_{2,0}}^2.$$

For the operator equation (5.3.3), we confirm now the form of the underlying positive definite operator equation (5.2.12), (5.2.13). The necessary conditions from minimizing the functional $LS(v,q)$ in (5.3.6) given by (5.2.9) read: find $(y,p) \in \mathcal{H}$ such that

$$(Ay + B'p, Av + B'q)_1 + (By, Bv)_2 = (f, Av + B'q)_1 + (u, Bv)_2 \tag{5.3.7}$$

for all $(v,q) \in \mathcal{H}$. Now let the Riesz operators $\mathcal{R}_1 : H'_{1,0} \to H_{1,0}$ and $\mathcal{R}_2 : H'_{2,0} \to H_{2,0}$ be defined as in (5.2.11),

$$\langle \mathcal{R}_i v, w \rangle = (v,w)_i, \quad v, w \in H'_{i,0}.$$

Then one can rewrite (5.3.7) in the form

$$\langle \mathcal{R}_1(Ay + B'p), Av + B'q \rangle + \langle \mathcal{R}_2 By, Bv \rangle = \langle \mathcal{R}_1 f, Av + B'q \rangle + \langle \mathcal{R}_2 u, Bv \rangle, \tag{5.3.8}$$

or, upon introducing

$$\hat{C}U := (A, B') \begin{pmatrix} y \\ p \end{pmatrix}, \quad U \in \mathcal{H}, \tag{5.3.9}$$

equivalently as

$$\langle \mathcal{R}_1 \hat{C}U, \hat{C}V \rangle + \langle \mathcal{R}_2 By, Bv \rangle = \langle \mathcal{R}_1 f, \hat{C}V \rangle + \langle \mathcal{R}_2 u, Bv \rangle.$$

This can in turn be written as

$$\langle \hat{C}' \mathcal{R}_1 \hat{C}U, V \rangle + \langle B' \mathcal{R}_2 By, v \rangle = \langle \hat{C}' \mathcal{R}_1 f, V \rangle + \langle B' \mathcal{R}_2 u, v \rangle.$$

Since this relation must be valid for all $V = (v,q)$, the resulting linear system reads

$$\hat{C}' \mathcal{R}_1 \hat{C} U + B' \mathcal{R}_2 B y = \hat{C}' \mathcal{R}_1 f + B' \mathcal{R}_2 u$$

or, inserting again (5.3.9),

$$\begin{pmatrix} A' \\ B \end{pmatrix} \mathcal{R}_1 (A, B') \begin{pmatrix} y \\ p \end{pmatrix} + B' \mathcal{R}_2 B y = \begin{pmatrix} A' \\ B \end{pmatrix} \mathcal{R}_1 f + B' \mathcal{R}_2 u.$$

In matrix–vector form, this is

$$\begin{pmatrix} A'\mathcal{R}_1A + B'\mathcal{R}_1B & A'\mathcal{R}_1A \\ B\mathcal{R}_1A & B\mathcal{R}_2B' \end{pmatrix} \begin{pmatrix} y \\ p \end{pmatrix} = \begin{pmatrix} A'\mathcal{R}_1 & B'\mathcal{R}_2 \\ B\mathcal{R}_1 & 0 \end{pmatrix} \begin{pmatrix} f \\ u \end{pmatrix}$$

or, equivalently,

$$\begin{pmatrix} A' & B' \\ B & 0 \end{pmatrix} \begin{pmatrix} \mathcal{R}_1 & \\ & \mathcal{R}_2 \end{pmatrix} \begin{pmatrix} A & B' \\ B & 0 \end{pmatrix} \begin{pmatrix} y \\ p \end{pmatrix} = \begin{pmatrix} A' & B' \\ B & 0 \end{pmatrix} \begin{pmatrix} \mathcal{R}_1 & \\ & \mathcal{R}_2 \end{pmatrix} \begin{pmatrix} f \\ u \end{pmatrix}. \tag{5.3.10}$$

Thus, we have established the following.

Remark 5.3 *For any saddle point problem of the form (5.3.3), the corresponding least squares problem is (5.3.10), where $\mathcal{R}_1$ and $\mathcal{R}_2$ are the Riesz operators defined by (5.2.11).*

We stress that the operator

$$\mathcal{M} := \begin{pmatrix} A' & B' \\ B & 0 \end{pmatrix} \begin{pmatrix} \mathcal{R}_1 & \\ & \mathcal{R}_2 \end{pmatrix} \begin{pmatrix} A & B' \\ B & 0 \end{pmatrix} \tag{5.3.11}$$

is symmetric and positive definite. In this notation, G in (5.2.12) is the right hand side in (5.3.10).

The particular example of interest is here the weak form of a second order elliptic problem of the type (4.2.1) which is treated by the Fictitious Domain—Lagrange Multiplier Approach (4.2.13) using $\square \supseteq \Omega$. In the notation from Chapter 2 and (4.2.10) with inner products from (4.2.11), (4.2.12), this problem is written in the following weak form: find $(y,p) \in H^1(\square) \times (H^{1/2}(\Gamma))'$ such that

$$\begin{aligned} A_1(v,(y,p)) &:= & a(y,v) + b(v,p) &= \langle f, v\rangle_\square, & v &\in H^1(\square), \\ A_2(q,(y,p)) &:= & b(y,q) &= \langle u, q\rangle_\Gamma, & q &\in (H^{1/2}(\Gamma))', \end{aligned} \tag{5.3.12}$$

is satisfied. Thus, (5.3.12) is of the form (5.3.2).

As seen in Section 4.2.4, the validity of the LBB condition constrains the mesh size of the trial spaces for the Lagrange multipliers in $(H^{1/2}(\Gamma))'$ relative to the mesh size on the domain $\square$. To facilitate a possibly stronger enforcement of the boundary conditions, a least squares formulation may offer an interesting alternative. Thus, to recast this in the setting of Section 5.2, we have here $H_1 = H_{1,0} = H^1(\square)$, B is the usual trace operator and therefore $\ker B' = \{0\}$, and $H_2 = H_{2,0} = (H^{1/2}(\Gamma))'$.

The relevant norms encountered in a least squares formulation (5.2.6) have in this case to be equivalent to

$$\|\cdot\|_{(H^1(\square))'} \quad \text{and} \quad \|\cdot\|_{H^{1/2}(\Gamma)}. \tag{5.3.13}$$

Another example that fits into the framework of Section 5.2 combines the above treatment of boundary conditions by Lagrange multipliers with a problem which by itself can already be formulated as a saddle point problem. In fact, the Stokes problem is an example for a vector–valued case with inhomogeneous velocity boundary conditions in addition to the divergence–free condition. Appending both side conditions by Lagrange multipliers leads to a three–by–three block saddle point problem which can then be placed into the above setting, see [DKS2].

5.4 Wavelet Representation of Least Squares Systems

Keeping the notation from Chapter 2 in mind, our substitute for the inner product $(\cdot,\cdot)_{H'_{i,0}}$ is of the form

$$(v,w)_i := \sum_{\lambda\in\mathbb{I}_i} d_{i,\lambda}^{-2}\,\langle v, {}^i\psi_\lambda\rangle\langle {}^i\psi_\lambda, w\rangle = \langle v, {}^i\Psi\rangle\, {}^i\mathbf{D}^{-2}\,\langle {}^i\Psi, w\rangle. \tag{5.4.1}$$

Remark 5.4 *From (2.18) and (2.22), we infer that*

$$C_{s_i}^{-2}\,(w,w)_i \le \|w\|^2_{H'_{i,0}} \le c_{s_i}^{-2}\,(w,w)_i, \tag{5.4.2}$$

so that the requirement (5.2.5) is satisfied.

In the following we will set for simplicity

$$c := \min_{1\le i\le M} c_{s_i}, \qquad C := \max_{1\le i\le M} C_{s_i}.$$

Note that the residuals $\mathcal{L}_iV - f_i$ are in (5.2.6) measured in norms equivalent to *dual* norms. Accordingly, the quantities in (5.4.1) involve coefficients like $\langle v, {}^i\psi_\lambda\rangle$ which are expansion coefficients in the dual bases ${}^i\Psi$. The exploitation of (2.22) for the evaluation of dual norms has been used before in several different contexts such as matrix compression [DPS2, DPS1], adaptive schemes [DDHS, CDD1] or stabilization [Be2, Be3].

In order to understand the nature of the scaling matrices ${}^i\mathbf{D}$ better, one has to specify the spaces $H_{i,0}$. Note that in all previous examples one has that the $H_{i,0}$ are subspaces of Sobolev spaces H^{s_i} on the respective domain. Therefore we will always assume that for some $s_i \in \mathbb{R}$, $i = 1,\dots,M$, one of the following situations occurs:

$$\begin{array}{llll} \text{(I)} & H_{i,0} \subseteq H^{s_i} \subseteq L_2, & & \text{i.e. } s_i \ge 0,\\ \text{(II)} & H'_{i,0} = H^{-s_i} & \text{when } H_{i,0} \supset L_2, & \text{i.e. } s_i < 0. \end{array} \tag{5.4.3}$$

Let us next express the inner products from (5.4.1) in terms of wavelet coefficients. Recall from (2.18) that ${}^i\mathbf{D}^{-1}\,{}^i\Psi$ is a Riesz basis in $H_{i,0}$. Thus, the expansion of $V \in \mathcal{H}$ has as in (2.25) the form

$$V = (\mathbf{v}_1^T\,({}^1\mathbf{D}^{-1})\,{}^1\Psi, \dots, \mathbf{v}_M^T\,({}^M\mathbf{D}^{-1})\,{}^M\Psi)^T. \tag{5.4.4}$$

As in Chapter 2, we catenate the individual bases ${}^i\Psi$, the corresponding coefficient sequences $\mathbf{v}_i$ and the scaling matrices ${}^i\mathbf{D}$ to

$$\Psi := \begin{pmatrix} {}^1\Psi \\ \vdots \\ {}^M\Psi \end{pmatrix}, \quad \mathbf{V} := \begin{pmatrix} \mathbf{v}_1 \\ \vdots \\ \mathbf{v}_M \end{pmatrix}, \quad \mathbf{D} := \operatorname{diag}\left({}^1\mathbf{D},\dots,{}^M\mathbf{D}\right). \tag{5.4.5}$$

Recall from (2.9) that

$$\langle {}^i\Psi, \mathcal{L}_iW\rangle = \sum_{l=1}^{M} A_{i,l}({}^i\Psi, {}^l\Psi)\,{}^l\mathbf{D}^{-1}\mathbf{w}_l. \tag{5.4.6}$$

Defining

$$\mathbf{A}^{i,l} := {}^i\mathbf{D}^{-1} A_{i,l}({}^i\Psi, {}^l\Psi)\, {}^l\mathbf{D}^{-1}, \tag{5.4.7}$$

(5.4.1) yields

$$(\mathcal{L}_i V, \mathcal{L}_i W)_i = \sum_{l,l'=1}^{M} \mathbf{v}_{l'}^T (\mathbf{A}^{i,l'})^T \mathbf{A}^{i,l} \mathbf{w}_l. \tag{5.4.8}$$

We summarize these observations as follows.

Remark 5.5 *For $(\cdot,\cdot)_i$ defined by (5.4.1) and the bilinear form $O(\cdot,\cdot)$ defined in (5.2.8),*

$$O(V, W) = \sum_{i=1}^{M} (\mathcal{L}_i V, \mathcal{L}_i W)_i,$$

one infers from (5.2.3) and (5.4.2) that

$$c_{\mathcal{L}} c \, \|V\|_{\mathcal{H}} \;\le\; O(V,V)^{1/2} \;\le\; C_{\mathcal{L}} C \, \|V\|_{\mathcal{H}}, \quad V \in \mathcal{H}. \tag{5.4.9}$$

Moreover, for $V, W \in \mathcal{H}$ expanded as in (5.4.4) one has

$$O(V, W) = \sum_{l,l'=1}^{M} \mathbf{v}_{l'}^T \left(\sum_{i=1}^{M} (\mathbf{A}^{i,l'})^T \mathbf{A}^{i,l} \right) \mathbf{w}_l = \mathbf{V}^T \mathbf{O} \mathbf{W}, \tag{5.4.10}$$

where

$$\mathbf{O} = \mathbf{A}^T \mathbf{A}, \quad \mathbf{A} := \left(\mathbf{A}^{i,l}\right)_{i,l=1}^{M}. \tag{5.4.11}$$

In particular, one therefore has

$$O(V, V) = \mathbf{V}^T \mathbf{O} \mathbf{V} = \|\mathbf{A}\mathbf{V}\|_{\ell_2(\mathbb{I})}^2. \tag{5.4.12}$$

The following result will play a pivotal role.

Theorem 5.6 *Consider in analogy to (5.4.5) the catenated right hand side data*

$$\mathbf{F} := \mathbf{D}^{-1}(\langle {}^1\Psi, f_1\rangle^T, \ldots, \langle {}^M\Psi, f_M\rangle^T)^T.$$

Then $U = \mathbf{U}^T \mathbf{D}^{-1} \Psi$ solves (5.2.9) if and only if $\mathbf{U}$ solves

$$\mathbf{O}\mathbf{U} \equiv \mathbf{A}^T \mathbf{A}\, \mathbf{U} = \mathbf{A}^T \mathbf{F}. \tag{5.4.13}$$

Moreover, the matrices $\mathbf{A}$, $\mathbf{O}$ define automorphisms of $\ell_2(\mathbb{I})$, where $\mathbb{I} = \mathbb{I}_1 \times \cdots \times \mathbb{I}_M$, i.e., in terms of the constants from (5.2.3) and (2.18), (5.4.2) one has

$$c^2 c_{\mathcal{L}} \, \|\mathbf{V}\|_{\ell_2(\mathbb{I})} \;\le\; \|\mathbf{A}\mathbf{V}\|_{\ell_2(\mathbb{I})} \;\le\; C^2 C_{\mathcal{L}} \, \|\mathbf{V}\|_{\ell_2(\mathbb{I})}. \tag{5.4.14}$$

Proof: By (5.2.8) and arguments quite analogous to the above derivations, one obtains for $V \in \mathcal{H}$

$$\begin{aligned} F(V) &= \sum_{i=1}^{M} (\mathcal{L}_i V, f_i)_i \\ &= \sum_{i=1}^{M} \langle \mathcal{L}_i V, {}^i\Psi \rangle \, ({}^i\mathbf{D}^{-2}) \, \langle {}^i\Psi, f_i \rangle \\ &= \sum_{l=1}^{M} \mathbf{v}_l^T \sum_{i=1}^{M} (\mathbf{A}^{i,l})^T \, ({}^i\mathbf{D}^{-1}) \, \langle {}^i\Psi, f_i \rangle \\ &= \mathbf{V}^T \mathbf{A}^T \mathbf{F}. \end{aligned}$$

Combining this with (5.4.10) shows that (5.4.13) is equivalent to (5.2.9) and hence (5.2.2). This proves the first part of the assertion. As for the rest, note that by (5.4.10)

$$\|\mathbf{AV}\|^2_{\ell_2(\mathbb{I})} = O(V, V) = \sum_{i=1}^{M} (\mathcal{L}_i V, \mathcal{L}_i V)_i.$$

Since (2.18) means in the present notation that

$$c \, \|\mathbf{V}\|_{\ell_2(\mathbb{I})} \;\le\; \|V\|_{\mathcal{H}} \;\le\; C \, \|\mathbf{V}\|_{\ell_2(\mathbb{I})},$$

the assertion follows from (5.4.2) and (5.2.3). ∎

Note that **DAD** is the (unscaled) wavelet representation of $\mathcal{L}$ while $\mathbf{M} := \mathbf{DOD}$ is the wavelet representation of the operator $\mathcal{M}$ in (5.2.13). Moreover, the Riesz operators from (5.2.11) correspond essentially to the diagonal scaling by $\mathbf{D}^{-2}$ as detailed next. In terms of wavelet expansions, the Riesz operators are identified as follows. For simplicity we will suppress the index i for a moment. The inner product which substitutes $(\cdot,\cdot)_{H_0'}$ is here denoted by $(\cdot,\cdot)_*$. By (5.4.1), we have

$$(w, v)_* = \left\langle \sum_{\lambda \in \mathbb{I}} d_\lambda^{-2} \langle w, \psi_\lambda \rangle \psi_\lambda, v \right\rangle = \langle \langle w, \Psi \rangle \mathbf{D}^{-2} \Psi, v \rangle$$

so that

$$\mathcal{R}w = \langle w, \Psi \rangle \mathbf{D}^{-2} \Psi = \sum_{\lambda \in \mathbb{I}} d_\lambda^{-2} \langle w, \psi_\lambda \rangle \psi_\lambda. \tag{5.4.15}$$

Since this is an expansion in terms of the primal wavelet basis Ψ, it is reasonable to reexpress $\mathcal{R}$ in terms of the wavelet coefficients $\langle w, \tilde{\psi}_\lambda \rangle$. In fact, inserting the representation $w = \langle w, \tilde{\Psi} \rangle \Psi$ or $\Psi = \langle \Psi, \Psi \rangle \tilde{\Psi}$ into (5.4.15), yields

$$\mathcal{R}w = \langle w, \tilde{\Psi} \rangle \langle \Psi, \Psi \rangle \mathbf{D}^{-2} \Psi = \langle w, \Psi \rangle \mathbf{D}^{-2} \langle \Psi, \Psi \rangle \tilde{\Psi}. \tag{5.4.16}$$

Remark 5.7 *The collection Ψ is a Riesz basis for L_2 if and only if the mass matrix $\langle \Psi, \Psi \rangle$ satisfies*

$$from(5.2.11) \|\langle \Psi, \Psi \rangle\|, \; \|\langle \Psi, \Psi \rangle^{-1}\| < \infty. \tag{5.4.17}$$

In fact, one has

$$\langle \Psi, \Psi \rangle^{-1} = \langle \tilde{\Psi}, \tilde{\Psi} \rangle. \tag{5.4.18}$$

Proof: Recall from Theorem 3.2 that Ψ is a Riesz basis in L_2 if and only if $\tilde{\Psi}$ is a Riesz basis in L_2. Using biorthogonality (3.2.32) and expanding the elements of $\tilde{\Psi}$ in terms of Ψ provides $\mathbf{I} = \langle \Psi, \tilde{\Psi} \rangle = \langle \Psi, \langle \tilde{\Psi}, \tilde{\Psi} \rangle \Psi \rangle = \langle \Psi, \Psi \rangle \langle \tilde{\Psi}, \tilde{\Psi} \rangle$. ■

In the same way one shows that

$$(w, v)_* = \langle \mathcal{R}w, v \rangle = \langle w, \mathcal{R}v \rangle, \tag{5.4.19}$$

i.e., $\mathcal{R}$ is self–adjoint. We shall later also employ the inverse of $\mathcal{R}$ which is given by

$$\mathcal{R}^{-1}v = \langle v, \tilde{\Psi} \rangle \mathbf{D}^2 \langle \tilde{\Psi}, \tilde{\Psi} \rangle \Psi. \tag{5.4.20}$$

Using biorthogonality (3.2.32) and the representation (5.4.18), it is straightforward to confirm (5.4.20) by applying $\mathcal{R}$ to the right hand side of (5.4.20).

We conclude this section with a simple observation which will be useful later.

Remark 5.8 *Suppose that Ψ is a Riesz basis in L_2 and that $v \in L_2$ satisfies*

$$\langle v, \tilde{\Psi} \rangle \mathbf{D}^2 \in \ell_2(\mathbb{I}). \tag{5.4.21}$$

Then one has $\mathcal{R}^{-1}v \in L_2$.

Proof: By Remark 5.7, $\langle v, \tilde{\Psi} \rangle \mathbf{D}^2 \in \ell_2(\mathbb{I})$ if and only if $\langle v, \tilde{\Psi} \rangle \mathbf{D}^2 \langle \tilde{\Psi}, \tilde{\Psi} \rangle \in \ell_2(\mathbb{I})$ so that the assertion follows from (5.4.20). ■

In fact, (5.4.21) is a regularity assumption under suitable assumptions on the underlying wavelet bases.

Theorem 5.6 states that the original variational problem (5.2.9) has been transformed into an equivalent infinite system of linear equations which is now well posed in Euclidean metric $\ell_2(\mathbb{I})$, completing Step 2 in the general concept from Chapter 2. Moreover, the already discrete nature of (5.4.13) suggests natural finite dimensional discretizations. To describe this, consider any finite set $\mathbf{\Lambda} = \Lambda_1 \times \cdots \times \Lambda_M \subset \mathbb{I}$ and let $\boldsymbol{\lambda} = (\lambda_1, \ldots, \lambda_M)$ be a typical element of $\mathbf{\Lambda}$. Let us denote by

$$\mathcal{S}_{\mathbf{\Lambda}} = S_{\Lambda_1} \times \cdots \times S_{\Lambda_M} \subset \mathcal{H} \tag{5.4.22}$$

a corresponding finite dimensional trial space. Obviously, a Galerkin scheme for (5.2.9) $O(U, V) = F(V)$ on $\mathcal{S}_{\mathbf{\Lambda}}$ would be equivalent to the linear system

$$\mathbf{O}_{\mathbf{\Lambda}} \mathbf{U}_{\mathbf{\Lambda}} = (\mathbf{A}^T \mathbf{F})|_{\mathbf{\Lambda}}, \tag{5.4.23}$$

where $\mathbf{O}_{\mathbf{\Lambda}} = (\mathbf{O}_{\boldsymbol{\lambda}, \boldsymbol{\lambda}'})_{\boldsymbol{\lambda}, \boldsymbol{\lambda}' \in \mathbf{\Lambda}}$ denotes the finite principal section of $\mathbf{O}$ determined by $\mathbf{\Lambda}$ and $(\mathbf{A}^T \mathbf{F})|_{\mathbf{\Lambda}}$ the restriction of the right hand side of (5.4.13) to $\mathbf{\Lambda}$. Theorem 5.6 would readily imply that the systems (5.4.23) would have spectral condition numbers bounded by

$$\kappa(\mathbf{O}_{\mathbf{\Lambda}}) \leq \frac{C_{\mathcal{L}}^2 C^4}{c_{\mathcal{L}}^2 c^4}$$

uniformly in $\mathbf{\Lambda}$. The problem is that neither the matrix $\mathbf{O_\Lambda}$ nor the right hand side $(\mathbf{A}^T\mathbf{F})|_{\mathbf{\Lambda}}$ can be computed exactly. In fact, denoting by ${}^i\mathbf{D}_{\Lambda_i}$ the 'restriction' of ${}^i\mathbf{D}$ to $\Lambda_i \subset I\!I_i$, the matrices

$$\mathbf{A}^{i,l}_{\hat{\Lambda}_i,\Lambda_l} := ({}^i\mathbf{D}^{-1}_{\hat{\Lambda}_i})\, A_{i,l}({}^i\Psi_{\hat{\Lambda}_i}, {}^l\Psi_{\Lambda_l})\, ({}^l\mathbf{D}^{-1}_{\Lambda_l}), \quad \mathbf{A}_{\hat{\mathbf{\Lambda}},\mathbf{\Lambda}} := \left(\mathbf{A}^{i,l}_{\hat{\Lambda}_i,\Lambda_l}\right)^M_{i,l=1} \tag{5.4.24}$$

are sections of the blocks from (5.4.7) and of the matrix $\mathbf{A}$ in (5.4.11), respectively. Clearly, the computation of the right hand side $(\mathbf{A}^T\mathbf{F})|_{\mathbf{\Lambda}}$ requires a multiplication of an infinite vector with the semi-infinite matrix $\mathbf{A}^T_{I\!I,\mathbf{\Lambda}}$ while $\mathbf{O_\Lambda} = \mathbf{A}^T_{I\!I,\mathbf{\Lambda}}\mathbf{A}_{I\!I,\mathbf{\Lambda}}$ is also a product of semi-infinite matrices. A natural way to obtain a *computable* problem is to truncate $\mathbf{A}_{I\!I,\mathbf{\Lambda}}$, i.e., to retain only finitely many rows corresponding to some finite index set $\hat{\mathbf{\Lambda}} \subset I\!I$ in the resulting *finite* matrix $\mathbf{A}_{\hat{\mathbf{\Lambda}},\mathbf{\Lambda}}$. This leads to the

Discrete Problem DP($\hat{\mathbf{\Lambda}}, \mathbf{\Lambda}$): *Given finite sets $\tilde{\mathbf{\Lambda}}, \mathbf{\Lambda} \subset I\!I$, set*

$$\mathbf{O}^{\hat{\mathbf{\Lambda}}}_{\mathbf{\Lambda}} := \mathbf{A}^T_{\hat{\mathbf{\Lambda}},\mathbf{\Lambda}}\mathbf{A}_{\hat{\mathbf{\Lambda}},\mathbf{\Lambda}} \tag{5.4.25}$$

and find $\mathbf{U_\Lambda} \in I\!R^{\mathbf{\Lambda}}$ such that

$$\mathbf{O}^{\hat{\mathbf{\Lambda}}}_{\mathbf{\Lambda}}\mathbf{U_\Lambda} = \mathbf{A}^T_{\hat{\mathbf{\Lambda}},\mathbf{\Lambda}}\mathbf{F}_{\hat{\mathbf{\Lambda}}}, \tag{5.4.26}$$

where

$$\mathbf{F_\Lambda} := \mathbf{D}^{-1}_{\mathbf{\Lambda}} \begin{pmatrix} \langle {}^1\Psi_{\Lambda_1}, f_1\rangle \\ \vdots \\ \langle {}^M\Psi_{\Lambda_M}, f_M\rangle \end{pmatrix} =: \langle \mathbf{\Psi_\Lambda}, F\rangle, \tag{5.4.27}$$

and

$$\mathbf{\Psi_\Lambda} := \begin{pmatrix} {}^1\Psi_{\Lambda_1} \\ \vdots \\ {}^M\Psi_{\Lambda_M} \end{pmatrix}, \quad \mathbf{U_\Lambda} := \begin{pmatrix} (\mathbf{u}_1)_{\Lambda_1} \\ \vdots \\ (\mathbf{u}_M)_{\Lambda_M} \end{pmatrix}, \quad \mathbf{D_\Lambda} := \mathrm{diag}\left({}^1\mathbf{D}_{\Lambda_1}, \ldots, {}^M\mathbf{D}_{\Lambda_M}\right). \tag{5.4.28}$$

Of course, replacing $\mathbf{O_\Lambda}$ by the computable matrix $\mathbf{O}^{\hat{\mathbf{\Lambda}}}_{\mathbf{\Lambda}}$ changes the variational problem so that the question of stability rises again. That is, given $\mathbf{\Lambda}$ which determines the trial space, how does one have to choose $\hat{\mathbf{\Lambda}}$ so that (5.4.26) is 'stable' in the sense of (5.2.10)? How does the choice of $\hat{\mathbf{\Lambda}}$ affect accuracy?

To answer these questions for the present scope of problems, it is helpful to note that truncation of $\mathbf{A}_{I\!I,\mathbf{\Lambda}}$ is equivalent to truncating the new inner products (5.4.1). To describe this, let us denote for $\Lambda_i \subset I\!I_i$ by ${}^iP_{\Lambda_i}v := \langle v, {}^i\tilde{\Psi}_{\Lambda_i}\rangle {}^i\Psi_{\Lambda_i}$ the canonical truncation projector onto S_{Λ_i} associated with such a pair of biorthogonal bases ${}^i\Psi, {}^i\tilde{\Psi}$ whose adjoint is obviously given by $({}^iP_{\Lambda_i})'v := \langle v, {}^i\Psi_{\Lambda_i}\rangle {}^i\tilde{\Psi}_{\Lambda_i}$. Observe that for any finite $\Lambda_i \subset I\!I_i$ one has in these terms

$$\begin{aligned}
[v, w]_{\Lambda_i} &:= \sum_{\lambda\in\Lambda_i} d^{-2}_{i,\lambda}\langle v, {}^i\psi_\lambda\rangle\langle {}^i\psi_\lambda, w\rangle \\
&= (v, w)_i - \left((I - ({}^iP_{\Lambda_i})')v, (I - ({}^iP_{\Lambda_i})')w\right)_i
\end{aligned} \tag{5.4.29}$$

which involves quantities arising in Galerkin discretizations and can be evaluated. The same reasoning as before yields the following fact.

Remark 5.9 *Setting*

$$O^{\hat{\Lambda}}(V,W) := \sum_{i=1}^{M} [\mathcal{L}_i V, \mathcal{L}_i W]_{\hat{\Lambda}_i}, \quad F_{\hat{\Lambda}}(V) := \sum_{i=1}^{M} [\mathcal{L}_i V, f_i]_{\hat{\Lambda}_i}, \tag{5.4.30}$$

then $\mathbf{U}_\Lambda$ *solves (5.4.26) if and only if* $U_\Lambda := \mathbf{U}_\Lambda^T \mathbf{D}_\Lambda^{-1} \Psi_\Lambda \in \mathcal{S}_\Lambda$ *solves the variational problem*

$$O^{\hat{\Lambda}}(U_\Lambda, V) \; = \; F_{\hat{\Lambda}}(V) \quad \text{for all} \;\; V \in \mathcal{S}_\Lambda, \tag{5.4.31}$$

which, in turn, corresponds to minimizing the least squares functional

$$LS_{\hat{\Lambda}}(V) := \sum_{i=1}^{M} [\mathcal{L}_i V - f_i, \mathcal{L}_i V - f_i]_{\hat{\Lambda}_i}. \tag{5.4.32}$$

Moreover, for $V, W \in \mathcal{H}$*, respectively* $V_\Lambda, W_\Lambda \in \mathcal{S}_\Lambda$*, according to (5.4.10), one has*

$$O^{\hat{\Lambda}}(V,W) = \mathbf{V}^T \mathbf{O}^{\hat{\Lambda}} \mathbf{W}, \quad O^{\Lambda}(V_\Lambda, W_\Lambda) = \mathbf{V}_\Lambda^T \mathbf{O}_\Lambda^{\hat{\Lambda}} \mathbf{W}_\Lambda. \tag{5.4.33}$$

We say that $\mathrm{DP}(\hat{\Lambda}, \Lambda)$ is *stable* if

$$O^{\hat{\Lambda}}(V,V) \; \sim \; O(V,V) \;\; \text{for all} \;\; V \in \mathcal{S}_\Lambda. \tag{5.4.34}$$

Remark 5.10 *Note that, in view of (5.4.29), (5.4.30) and (5.4.1), one trivially has*

$$O^{\hat{\Lambda}}(V,V) \; \leq \; O(V,V), \quad V \in \mathcal{H}. \tag{5.4.35}$$

One expects that the truncation sets $\hat{\Lambda}_i$ in (5.4.29) should at least contain the sets Λ_i defining the trial spaces in order to ensure coercivity of the forms $O^{\hat{\Lambda}}(\cdot,\cdot)$ on $\mathcal{S}_\Lambda$. We will discuss this issue first from the view point of Galerkin discretizations as already suggested by the ingredients $\mathbf{A}^{i,l}_{\hat{\Lambda}_i,\Lambda_l}$. The next simple observation in this regard supports this interrelation.

Remark 5.11 *Suppose that* U_Λ *is a Galerkin solution in* $\mathcal{S}_\Lambda$*, i.e.,*

$$\langle \mathcal{L}_i U_\Lambda, v_i \rangle = \langle f_i, v_i \rangle, \quad v_i \in S_{\Lambda_i}, \; i = 1, \ldots, M. \tag{5.4.36}$$

Then $LS_\Lambda(U_\Lambda) = 0$*, i.e.,* U_Λ *solves* $DP(\Lambda, \Lambda)$ *and hence (5.2.9) for* $\hat{\Lambda} = \Lambda$*.*

Proof: By (5.4.29) and (5.4.36), one has for $i = 1, \ldots, M$,

$$[\mathcal{L}_i U_\Lambda - f_i, \mathcal{L}_i U_\Lambda - f_i]_{\Lambda_i} = \sum_{\lambda \in \Lambda_i} d_{i,\lambda}^{-2} \langle \mathcal{L}_i U_\Lambda - f_i, {}^i\psi_\lambda \rangle \langle {}^i\psi_\lambda, \mathcal{L}_i U_\Lambda - f_i \rangle = 0.$$

■

5.5 Truncation

Under certain circumstances $\mathrm{DP}(\mathbf{\Lambda}, \mathbf{\Lambda})$ is already stable, that means, the truncated forms $O^{\mathbf{\Lambda}}(\cdot,\cdot)$ are coercive over $\mathcal{S}_{\mathbf{\Lambda}}$ for arbitrary index sets $\mathbf{\Lambda}$ and corresponding trial spaces $\mathcal{S}_{\mathbf{\Lambda}}$. We refer to this as *symmetric truncation.*

Remark 5.12 *Assume that the Galerkin scheme for (5.2.2) is stable, i.e., defining*

$$\mathcal{P}_{\mathbf{\Lambda}} : \mathcal{H} \to \mathcal{S}_{\mathbf{\Lambda}} \quad by \quad \mathcal{P}_{\mathbf{\Lambda}} V := ({}^1P_{\Lambda_1} v_1, \ldots, {}^M P_{\Lambda_M} v_M)^T,$$

one has

$$\|\mathcal{P}'_{\mathbf{\Lambda}} \mathcal{L} V\|_{\mathcal{H}'} \sim \|V\|_{\mathcal{H}} \quad for\ all\ V \in \mathcal{S}_{\mathbf{\Lambda}}. \tag{5.5.1}$$

Then $DP(\mathbf{\Lambda}, \mathbf{\Lambda})$ is stable.

This assertion has been shown in [DKS2]. In addition one can infer that Galerkin schemes are stable if the bilinear form induced by $\mathcal{L}$ is coercive, i.e.,

$$\langle \mathcal{L}V, V \rangle := \sum_{i=1}^{M} A_i(V, v_i) \gtrsim \|V\|^2_{\mathcal{H}}. \tag{5.5.2}$$

This covers for instance the transmission problem considered in [DKS2].

Of course, (5.5.2) does not apply to the saddle point problem (5.3.3). There stability of the Galerkin scheme requires the validity of the LBB condition which was to be avoided.

Thus, we concentrate in the following on the question how to restore stability in the general setting. First of all, by (5.2.5), one has

$$\|(I - ({}^iP_{\Lambda_i})')\mathcal{L}_i V\|_i \lesssim \|\mathcal{L}_i V\|_i$$

so that

$$\sum_{i=1+r}^{M} \left((I - ({}^iP_{\Lambda_i})')\mathcal{L}_i V, (I - ({}^iP_{\Lambda_i})')\mathcal{L}_i V\right)_i \lesssim \sum_{i=1}^{M} \|\mathcal{L}_i V\|^2_{H'_{i,0}}. \tag{5.5.3}$$

One idea is to seek for possibly sharp upper bounds of the left hand side of (5.5.3) which can be numerically evaluated. This will generally depend on the particular case at hand. Therefore, we will focus here on an alternative strategy that works conceptually in the same way for all the different cases.

Recall that in the above notation $\mathbf{\Lambda}$ determines the trial spaces while $\hat{\mathbf{\Lambda}}$ defines the variational problem (5.4.26). It is clear that the larger $\hat{\mathbf{\Lambda}}$ relative to a given $\mathbf{\Lambda}$ is, the better $\mathbf{A}_{\hat{\mathbf{\Lambda}},\mathbf{\Lambda}}$ and $\mathbf{O}^{\hat{\mathbf{\Lambda}}}_{\mathbf{\Lambda}}$ approximate $\mathbf{A}_{\mathbb{I},\mathbf{\Lambda}}$, respectively the ideal matrix $\mathbf{O}_{\mathbf{\Lambda}}$ from (5.4.23). We will show that $\#\hat{\mathbf{\Lambda}}$ can always be kept proportional to $\mathbf{\Lambda}$ uniformly in $\mathbf{\Lambda}$ to ensure stability of $\mathrm{DP}(\hat{\mathbf{\Lambda}}, \mathbf{\Lambda})$ in (5.4.26).

To this end, a slightly sharper version of (5.5.3) is needed to bound $O^{\hat{\mathbf{\Lambda}}}(\cdot,\cdot)$ from below by $O(\cdot,\cdot)$ on $\mathcal{S}_{\mathbf{\Lambda}}$. We will refer to this as *expanded truncation.*

It is instructive to consider first the case where the index sets $\Lambda_i, \hat{\Lambda}_i$ reflect *uniform refinements.* To simplify notation we choose the same difference in the highest refinement level for all components although this is by no means essential. Thus, denote by J_i and $J_i + \hat{L}$ the finest level in Λ_i and $\hat{\Lambda}_i$, respectively. We discuss in the following the choice of $\hat{L} \geq 0$.

Proposition 5.13 *Let the wavelet bases be chosen so that*

$$\begin{aligned} t_i &> s_i && \text{for (5.4.3)(I)}, \\ \tilde{t}_i &> -s_i && \text{for (5.4.3)(II)}, \end{aligned} \tag{5.5.4}$$

where the t_i and $\tilde{t}_i$ are the regularity bounds from (3.2.57). Furthermore, let

$$\begin{aligned} 0 &< \varepsilon < t_i - s_i && \text{in the case (I)}, \\ 0 &< \varepsilon < \tilde{t}_i + s_i && \text{in the case (II)}, \end{aligned}$$

and assume that

$$\|\mathcal{L}_i V\|^2_{H_i^{-s_i+\varepsilon}} \lesssim \sum_{i=1}^{M} \|v_i\|^2_{H_i^{s_i+\varepsilon}}, \quad V \in \mathcal{H}^{+\varepsilon} := \prod_{i=1}^{M} H_i^{s_i+\varepsilon}. \tag{5.5.5}$$

Then there exists a fixed positive integer $\hat{L}$ such that one has for any $\mathbf{J} = (J_1, \ldots, J_M) \in \mathbb{N}^M$

$$O^{\mathbf{J}+\hat{L}}(V, V) \sim \|V\|^2_{\mathcal{H}}, \quad V \in \mathcal{S}_{\mathbf{J}} := \mathcal{S}_{\boldsymbol{\Lambda}}, \tag{5.5.6}$$

where $\mathbf{J} + \hat{L} := J_1 + \hat{L}, \ldots, J_M + \hat{L}$.

Here H_i^s are the Sobolev spaces H_i^s as in Section 3.2. Condition (5.5.5) means that $\mathcal{L}$ is assumed to be still bounded in spaces of somewhat higher regularity.

Proof: Let us abbreviate ${}^iP_{J_i+\hat{L}} := {}^iP_{\hat{\Lambda}_i}$. Consider first the case (5.4.3)(I). One infers from (3.2.61) that the $({}^iP_{J_i+\hat{L}})'$ are uniformly bounded in $H'_{i,0}$. Combining a standard duality and interpolation argument based on the direct estimates (3.2.55), one obtains for any $V \in \mathcal{S}_{\mathbf{J}}$,

$$\begin{aligned} \|(I - ({}^iP_{J_i+\hat{L}})')\mathcal{L}_i V\|^2_{H'_{i,0}} &\lesssim 2^{-2\varepsilon(J_i+\hat{L})}\|\mathcal{L}_i V\|^2_{H_i^{-s_i+\varepsilon}} \\ &\lesssim 2^{-2\varepsilon(J_i+\hat{L})}\sum_{i=1}^{M}\|v_i\|^2_{H_i^{s_i+\varepsilon}}, \end{aligned} \tag{5.5.7}$$

where we have used (5.5.5) in the last step. Applying now the inverse estimate (3.2.56) to each summand on the right hand side of (5.5.7), one concludes that

$$\|(I - ({}^iP_{J_i+\hat{L}})')\mathcal{L}_i V\|^2_{H'_{i,0}} \lesssim 2^{-2\varepsilon(J_i+\hat{L})}2^{2\varepsilon J_i}\|V\|^2_{\mathcal{H}}, \quad V \in \mathcal{S}_{\mathbf{J}}. \tag{5.5.8}$$

Hence, keeping (5.2.5) in mind, there exists a constant C independent of $\mathbf{J}, \hat{L}$ such that

$$\sum_{i=1}^{M}\Big((I - ({}^iP_{J_i+\hat{L}})')\mathcal{L}_i V, (I - P'_{i,J_i+\hat{L}})\mathcal{L}_i V\Big)_i \leq C\,2^{-2\varepsilon\hat{L}}\|V\|^2_{\mathcal{H}}, \quad V \in \mathcal{S}_{\mathbf{J}}. \tag{5.5.9}$$

Choosing $\hat{L}$ large enough so that $C\,2^{-2\hat{L}\varepsilon} \leq 1/2$, the assertion follows from (5.4.9), (5.4.29) and (5.4.30). ∎

Thus, when dealing with uniformly refined trial spaces, one can always choose a truncation level depending on the current trial spaces which yields a computable coercive energy inner product $O^{\mathbf{J}+\hat{L}}(\cdot,\cdot)$. Clearly, since $\hat{L}$ is fixed, the number of terms appearing in $O^{\mathbf{J}+\hat{L}}(\cdot,\cdot)$ stays proportional to $\dim \mathcal{S}_{\mathbf{J}}$.

The argument is more involved when the index sets $\mathbf{\Lambda}$ are *nonuniform*. In order to estimate the truncation effect in (5.4.7) in the general case, we return to the representation of $O^{\hat{\mathbf{\Lambda}}}(\cdot,\cdot)$ in wavelet coordinates (5.4.33).

One obtains for $V \in \mathcal{S}_{\mathbf{\Lambda}}$ given by (5.4.4)

$$((I - ({}^iP_{\hat{\Lambda}_i})')\mathcal{L}_i V, (I - ({}^iP_{\hat{\Lambda}_i})')\mathcal{L}_i V)_i = \sum_{l=1}^{M} \mathbf{v}_l^T (\mathbf{A}^{i,l}_{I\!I_i \setminus \hat{\Lambda}_i, \Lambda_l})^T \mathbf{A}^{i,l}_{I\!I_i \setminus \hat{\Lambda}_i, \Lambda_l} \mathbf{v}_l. \qquad (5.5.10)$$

The point is now that for *all* operators arising in the examples in [DKS2] including the integral operators and the problems considered here, these matrices are *nearly sparse* in a sense to be explained next. In fact, the $\mathcal{L}_{i,l}$ are either *local*, i.e.,

$$\langle {}^i\psi_\nu, \mathcal{L}_{i,l}\, {}^l\psi_\lambda \rangle = A_{i,l}({}^i\psi_\nu, {}^l\psi_\lambda) = 0 \quad \text{if } \operatorname{supp} {}^l\psi_\lambda \cap \operatorname{supp} {}^i\psi_\nu = \emptyset, \qquad (5.5.11)$$

like for trace or differential operators covering the above specific saddle point (4.2.13), or they have a global Schwartz kernel

$$(\mathcal{L}_{i,j} v)(x) = \int K_{i,j}(x,y) v(y) dy$$

with the Calderón-Zygmund property

$$\left|\partial_x^\alpha \partial_y^\beta K_{i,l}(x,y)\right| \lesssim \operatorname{dist}(x,y)^{-(d_i + 2\tau_{i,l} + |\alpha| + |\beta|)}, \qquad (5.5.12)$$

when $2\tau_{i,l}$ is the order of the operator $\mathcal{L}_{i,l}$ and d_i is the spatial dimension of the domain to which the space H_i refers. Of course, by assumption (5.2.4), one always has

$$|\tau_{i,l}| \leq |s_i|, \quad \text{when } H_{i,0} \subseteq H^{s_i}. \qquad (5.5.13)$$

Under these assumptions one can show that the matrices $\mathbf{A}^{i,l}$ exhibit a decay away from the diagonal. To quantify this, it is helpful to revisit the Fictitious Domain–Lagrange Multiplier Problem (5.3.12).

In fact, the previously considered problem in weak formulation (5.3.12) exhibits typical situations which we will briefly discuss now. Recall that the wavelet basis for $H_1 = H^1(\square)$ is denoted by ${}^1\Psi$. Moreover, one has here $d_{1,\lambda} = 2^{|\lambda|}$ so that

$$(v,w)_1 = \sum_{\lambda \in I\!I_1} \langle v, {}^1\psi_\lambda \rangle\, 2^{-2|\lambda|} \langle {}^1\psi_\lambda, w \rangle.$$

The regularity bounds for these wavelets should satisfy ${}^1t = t({}^1\Psi) > 1$. In order to analyze the behavior of $\mathbf{A}^{1,1} = {}^1\mathbf{D}^{-1} \langle \nabla {}^1\Psi, \nabla {}^1\Psi \rangle\, {}^1\mathbf{D}^{-1}$, it suffices to consider a typical entry

$$2^{-(|\lambda|+|\lambda'|)} |\langle \nabla {}^1\psi_\lambda, \nabla {}^1\psi_{\lambda'} \rangle|.$$

Bounds for these quantities are meanwhile standard, and a crude estimate yields

$$|2^{-(|\lambda|+|\lambda'|)}|\,|\langle\nabla\, {}^1\psi_\lambda, \nabla\, {}^1\psi_{\lambda'}\rangle| \lesssim \begin{cases} 2^{(\sigma'+d_1/2)||\lambda|-|\lambda'||}, & \text{if } {}^1\Omega_\lambda \cap {}^1\Omega_{\lambda'} \neq \emptyset, \\ 0, & \text{if } {}^1\Omega_\lambda \cap {}^1\Omega_{\lambda'} = \emptyset, \end{cases} \tag{5.5.14}$$

where ${}^i\Omega_\lambda := \operatorname{supp} {}^i\psi_\lambda$ and σ' is the degree of Hölder continuity of $\nabla\, {}^1\psi_\lambda$, see e.g. [DDHS]. Thus, aside from spatial location the entries in $\mathbf{A}^{1,1}$ exhibit an exponential decay in scale whose strength is determined by the regularity of the wavelets.

The operator $\mathcal{L}_{1,2}$ is essentially the adjoint of the trace operator. Recall that ${}^2\Psi$ are wavelets living on Γ. Furthermore, ${}^2\mathbf{D} = \operatorname{diag}(2^{-|\lambda|/2} : \lambda \in I\!I_2)$. Thus, one has

$$\begin{aligned}(\mathbf{A}^{1,2})_{\lambda,\lambda'} &= 2^{-|\lambda|}\,\langle {}^1\psi_\lambda, B'\, {}^2\psi_{\lambda'}\rangle_\square\, 2^{|\lambda'|/2} \\ &= 2^{-|\lambda|+|\lambda'|/2}\,\langle B\, {}^1\psi_\lambda, {}^2\psi_{\lambda'}\rangle_\Gamma.\end{aligned} \tag{5.5.15}$$

Assume first that $|\lambda'| > |\lambda|$ and apply Hölder's inequality to conclude that

$$\begin{aligned}\langle B\, {}^1\psi_\lambda, {}^2\psi_{\lambda'}\rangle_\Gamma &\leq \|B\, {}^1\psi_\lambda\|_{H^{\sigma+1/2}(\Gamma)}\, \|{}^2\psi_{\lambda'}\|_{H^{-\sigma-1/2}(\Gamma)} \\ &\lesssim \|{}^1\psi_\lambda\|_{H^{\sigma+1}(\square)}\, \|{}^2\psi_{\lambda'}\|_{H^{-\sigma-1/2}(\Gamma)} \\ &\lesssim 2^{(1+\sigma)|\lambda|}\, 2^{-(\sigma+1/2)|\lambda'|},\end{aligned} \tag{5.5.16}$$

where we have used the trace theorem Corollary 4.10 and the norm equivalences (3.2.61). Note that for the application of the latter norm equivalence σ is subject to constraints depending on the choice of wavelet bases, namely

$$-t({}^1\tilde{\Psi}) < \pm\sigma + 1/2 < -t({}^1\Psi), \quad -t({}^2\tilde{\Psi}) < \pm\sigma - 1/2 < -t({}^2\Psi), \tag{5.5.17}$$

and the fact that the involved operators are bounded in some range above the respective energy spaces. Combining (5.5.16) with (5.5.15) yields for σ satisfying (5.5.17)

$$|(\mathbf{A}^{1,2})_{\lambda,\lambda'}| \lesssim \begin{cases} 2^{\sigma||\lambda|-|\lambda'||}, & \text{if } {}^1\Omega_\lambda \cap {}^2\Omega_{\lambda'} \neq \emptyset, \\ 0, & \text{if } {}^1\Omega_\lambda \cap {}^2\Omega_{\lambda'} = \emptyset, \end{cases} \tag{5.5.18}$$

The other two blocks appearing in $\mathbf{A}$ are either zero or the transpose of $\mathbf{A}^{1,2}$.

In summary, we see that the block matrices $\mathbf{A}^{i,l}$ from (5.4.7) satisfy estimates of the type

$$|(\mathbf{A}^{i,l})_{\lambda,\lambda'}| \lesssim \frac{2^{-||\lambda|-|\lambda'||\sigma_{i,l}}}{(1 + 2^{\min(|\lambda|,|\lambda'|)}\operatorname{dist}({}^i\Omega_\lambda, {}^l\Omega_{\lambda'}))^{\beta_{i,l}}}, \tag{5.5.19}$$

where $\sigma_{i,l} > n_i/2$ depends on the regularity of the wavelets ${}^i\Psi, {}^l\Psi$, n_i being the spatial dimension, and $\beta_{i,l} > d_i$ depends on the order $\tau_{i,l}$ of the operator $\mathcal{L}_{i,l}$ and on the order of cancellation properties $\tilde{d}_i, \tilde{d}_l$ of the wavelets ${}^i\Psi, {}^l\Psi$.

Remark 5.14 *We have seen that the saddle point problem (4.2.13) involves only differential and trace operators which are both local from which one can derive (5.5.19). In the case of arbitrary saddle point problems of the form (5.3.3) involving non–local operators, one has to assure that decay estimates (5.5.19) holds, see also [Sch].*

With the estimates (5.5.19) at hand, we can now follow [DDHS] to construct for a given index set $\mathbf{\Lambda}$ for the trial spaces $\mathcal{S}_{\mathbf{\Lambda}}$ a sufficiently larger index set $\hat{\mathbf{\Lambda}}$ for the truncation in (5.4.7). Here it is important to have $\sigma_{i,l} > n_i/2$. In fact, for $\sigma_{i,l}$ as above let $0 < \delta < \sigma_{i,l} - n_i/2$. For a given $\eta > 0$ choose positive numbers η_1, η_2 such that

$$\eta_1^{\beta_{i,l}-n_i} + 2^{-\delta/\eta_2} \leq \eta, \tag{5.5.20}$$

and define for $\lambda \in \mathbb{I}_i$

$$\mathbb{I}_{\lambda,\eta} := \left\{\lambda' \in \mathbb{I}_l : ||\lambda| - |\lambda'|| \leq \eta_2^{-1} \text{ and } 2^{\min\{|\lambda|,|\lambda'|\}}\text{dist}\,({}^i\Omega_\lambda, {}^l\Omega_{\lambda'}) \leq \eta_1^{-1}\right\}. \tag{5.5.21}$$

Setting

$$\hat{\Lambda}_i := \bigcup_{\lambda\in\Lambda_i} \mathbb{I}_{\lambda,\eta}, \tag{5.5.22}$$

it has been shown in [DDHS] that

$$\|\mathbf{A}^{i,l}_{\mathbb{I}_i\setminus\hat{\Lambda}_i,\ \Lambda_l}\| \leq \hat{C}\eta, \tag{5.5.23}$$

where the constant $\hat{C}$ depends on $\sigma_{i,l}, \delta, \tau_{i,l}$ and the constants in (2.22).

Proposition 5.15 *Suppose that the blocks $\mathbf{A}^{i,l}$ satisfy (5.5.19), e.g., under the above assumptions on $\mathcal{L}$. Then for any fixed $\alpha \in (0,1)$ and any $\mathbf{\Lambda}$ in (5.4.22) one can choose some $\hat{\mathbf{\Lambda}}$ such that*

$$\alpha O(V,V) \ \leq\ O^{\hat{\mathbf{\Lambda}}}(V,V) \ \leq\ O(V,V), \quad V \in \mathcal{S}_{\mathbf{\Lambda}}, \tag{5.5.24}$$

and

$$\#\hat{\mathbf{\Lambda}} \ \leq\ c_\#\#\mathbf{\Lambda},$$

where the constant $c_\#$ depends on the constants $c, C, c_{\mathcal{L}}, O_{\mathcal{L}}, \alpha$ but not on $\mathbf{\Lambda}$ and V.

Proof: We infer from (5.5.10) and (5.5.23) that

$$\begin{aligned}
\sum_{i=1}^{M}\left((I-({}^iP_{\Lambda_i})')\mathcal{L}_iV,(I-({}^iP_{\Lambda_i})')\mathcal{L}_iV\right)_i &= \sum_{i=1}^{M}\sum_{l=1}^{M}\mathbf{v}_l^T(\mathbf{A}^{i,l}_{\mathbb{I}_i\setminus\hat{\Lambda}_i,\Lambda_l})^T\mathbf{A}^{i,l}_{\mathbb{I}_i\setminus\hat{\Lambda}_i,\Lambda_l}\mathbf{v}_l \\
&\leq (\hat{C}\eta)^2\sum_{i=1}^{M}\sum_{l=1}^{M}\|\mathbf{v}_l\|^2_{\ell_2(\Lambda_l)} \qquad (5.5.25)\\
&\leq (c^2c_{\mathcal{L}})^{-1}M(\hat{C}\eta)^2\sum_{l=1}^{M}(\mathcal{L}_lV,\mathcal{L}_lV)_l,
\end{aligned}$$

where we have used (5.2.8), (5.4.12) and (5.4.14) in the last step. Now choose η so that $M(c^2c_{\mathcal{L}})^{-1}(\hat{C}\eta)^2 < 1-\alpha$. Keeping (5.4.35) in mind, the first relation in (5.5.24) follows now from (5.4.7) and (5.4.30). Moreover, once η is fixed, the cardinality of $\mathbb{I}_{\lambda,\eta}$ remains uniformly bounded in λ. This completes the proof. ∎

We hasten to add though that the above estimate is rather crude and should only demonstrate the principal accessibility of the involved quantities. In fact, a somewhat different point of view is more adequate which is motivated by the developments in [CDD1]. From this point of view one would aim at *adaptively evaluating* for given $\mathbf{v}_l \in \ell_2(\Lambda_l)$ the matrix/vector products $\mathbf{A}^{i,l}_{\mathbb{I}_i,\Lambda_l}\mathbf{v}$ with a certain precision required by the current state of the adaptive solver.

5.6 Preconditioning and Computational Work

To estimate the total amount of work needed for solving (5.4.26), suppose first that all the operators in $\mathcal{L}$ are local (5.5.11) and that all index sets in $\mathbf{\Lambda}$ in (5.4.22) correspond to uniform refinements where $J_i \in \mathbb{N}$ denotes the highest level in Λ_i. In all the mentioned constructions of wavelet bases one can exhibit multiscale transformations of the form

$$ {}^i\Psi^{J_i} = {}^i\mathbf{T}_{J_i}{}^i\Phi_{J_i}, \quad i = 1, \ldots, M, \tag{5.6.1}$$

compare (3.2.22), where the ${}^i\Phi_{J_i} = \{{}^i\phi_{J_i,k} : k \in \Delta_{J_i}\}$ are the single scale bases for the respective multiresolution spaces ${}^iS_{J_i} = S({}^i\Psi^{J_i})$, see (3.2.3) and (3.2.49). As specified in Section 3.4, the ${}^i\Phi_{J_i}$ consist typically of parametric liftings of tensor product B–splines so that $\operatorname{diam}(\operatorname{supp}{}^i\phi_{J_i,k}) \sim 2^{-J_i}$, cf. (3.2.5). They play the role of nodal bases on the highest refinement level. Moreover, the transformations ${}^i\mathbf{T}_{J_i}$ interrelating single and multiscale bases are fast in the sense that the application ${}^i\mathbf{T}_{J_i}\mathbf{c}$ requires the order of $\#\Delta_{J_i}$ operations, see Remark 3.1. Now a typical block $\mathbf{A}^{i,l}_{J_i+\hat{L},J_l}$ (5.4.24) in $\mathbf{A}_{\mathbf{J}+\hat{L},\mathbf{J}}$ can be written as

$$ \mathbf{A}^{i,l}_{J_i+\hat{L},J_l} = {}^i\mathbf{D}^{-1}_{J_i+\hat{L}}\,{}^i\mathbf{T}^T_{J_i+\hat{L}}\,A_{i,l}({}^i\Phi_{J_i+\hat{L}}, {}^l\Phi_{J_l})\,{}^l\mathbf{T}_{J_l}\,{}^l\mathbf{D}^{-1}_{J_l}, \tag{5.6.2}$$

where the blocks $A_{i,l}({}^i\Phi_{J_i+\hat{L}}, {}^l\Phi_{J_l})$ are in view of the locality of the single scale bases (3.2.5) uniformly sparse. By the above property of the transformations ${}^i\mathbf{T}_{J_i}$, an application of $\mathbf{O}^{\mathbf{J}+\hat{L}}_{\mathbf{J}}$ in (5.4.26) to a vector still requires a number of arithmetic operations which stays proportional to the size of $\mathbf{O}^{\mathbf{J}+\hat{L}}_{\mathbf{J}}$.

We recall next that the isomorphisms (2.18) and (2.22) not only allow one to access the dual norms but also provide asymptotically optimal preconditioners once $\mathrm{DP}(\hat{\mathbf{\Lambda}}, \mathbf{\Lambda})$ has been arranged to be stable. Combining Proposition 5.15 with Theorem 5.6 yields the following fact.

Theorem 5.16 *Given $\mathbf{\Lambda}$, one can find $\hat{\mathbf{\Lambda}}$ satisfying*

$$ \#\hat{\mathbf{\Lambda}} \sim \#\mathbf{\Lambda} $$

such that $\mathrm{DP}(\hat{\mathbf{\Lambda}}, \mathbf{\Lambda})$ *is stable and the matrices $\mathbf{O}^{\hat{\mathbf{\Lambda}}}_{\mathbf{\Lambda}}$ in (5.4.26) are symmetric positive definite and have uniformly bounded spectral condition numbers*

$$ \kappa(\mathbf{O}^{\hat{\mathbf{\Lambda}}}_{\mathbf{\Lambda}}) \le \alpha^{-1}\kappa(\mathbf{O}_{\mathbf{\Lambda}}) \le \frac{C^2_{\mathcal{L}}C^4}{\alpha c^2_{\mathcal{L}}c^4}, \tag{5.6.3}$$

where α is the constant from (5.5.24).

Thus, for a proper expanded truncation and scaling (5.4.4) one automatically obtains symmetric positive definite linear systems with uniformly bounded condition numbers.

As for estimating the amount of operations using a conjugate gradient method for the solution of (5.4.26), let $\mathbf{J}$ denote some desired highest discretization level. Since the matrices $\mathbf{O}_{\Lambda}^{\hat{\Lambda}} = \mathbf{O}_{\mathbf{J}}^{\mathbf{J}+\hat{L}}$ have uniformly bounded condition numbers, in each iteration the error is reduced by a fixed amount. Let $N_{\mathbf{J}}$ be the number of unknowns and let $m_{\mathbf{J}}$ be the amount of iterations which is needed to reduce the initial error below a fixed prescribed accuracy ε. Thus, one needs $\mathcal{O}(\log \varepsilon)$ iterations. Since each iteration requires in view of the uniformly bounded condition numbers $\mathcal{O}(N_{\mathbf{J}})$ operations, one therefore needs a total amount of $\mathcal{O}(N_{\mathbf{J}} \log \varepsilon)$ arithmetic operations.

One can further get rid of the logarithmic factor by using *nested iteration* in combination with the conjugate gradient method. Given an ascending sequence of trial spaces, one can use the approximate solution on a given level as initial guess for the next higher level. By Theorem 5.16, only a uniformly bounded number of conjugate gradient iterations is required to reduce the error by a fixed fraction. This is all that is needed when realizing on each level discretization error accuracy. Thus, on each refinement level a uniformly bounded number of matrix/vector multiplications suffices to solve the discrete problem with an accuracy comparable to the truncation error. Hence, employing a sufficient increase of degrees of freedom when progressing to higher levels as e.g. in the case of uniform refinements, a standard geometric series argument shows that the overall work stays proportional to the computational work required by a matrix/vector multiplication on the highest level $\mathbf{J}$, that is, the work is proportional to $\mathcal{O}(N_{\mathbf{J}})$. Summarizing this, we have proved the following.

Proposition 5.17 *Prescribing some finest discretization level* $\mathbf{J}$*, the system (5.4.26) can be solved up to truncation error accuracy by a conjugate gradient method combined with nested iteration with an overall amount of work* $\mathcal{O}(N_{\mathbf{J}})$ *where* $N_{\mathbf{J}}$ *is the number of unknowns on level* $\mathbf{J}$*.*

5.7 Numerical Experiments

We choose the previously discussed example for the numerical experiments because it involves two representative 'difficult' norms, namely, the H^{-1}-norm and a fractional trace norm. As the fictitious domain, we take again $\square = (0,1)^2$ containing Ω as a disc with radius 1/2 from (4.3.2) and consider the elliptic boundary value problem (4.3.1),

$$\begin{aligned} -\Delta y + y &= 1 \quad \text{in } \Omega, \\ y &= 0 \quad \text{on } \Gamma = \partial\Omega. \end{aligned}$$

Appending the boundary conditions by Lagrange multipliers yields the weak formulation (4.2.10). Recall from Section 5.3 that in this case one has $H_{1,0} = H^1(\square)$, $H_{2,0} = H^{-1/2}(\Gamma)$. As before we have deliberately chosen here the circle as a boundary to test the effect of a boundary which is *not* aligned with the coordinate axes.

We use $\ell = J_2$ to denote the refinement level on the boundary Γ while $j = J_1$ refers always to the domain $\square$, i.e., $\mathbf{J} = (j, \ell)$. Since the wavelet bases on the boundary

j	ℓ	$\hat{L}$	#it	$\|\|\mathbf{y}_\Lambda\|_\Gamma\|\|_{\ell_2}$	tol
4	3	0	25	$2.679e-02$	$5.625e-01$
4	4	0	25	$5.085e-02$	$5.625e-01$
4	5	0	25	$7.764e-02$	$5.625e-01$
4	6	0	25	$2.275e-01$	$5.625e-01$
4	7	0	25	$2.989e-01$	$5.625e-01$
4	8	0	25	$3.106e-01$	$5.625e-01$
5	3	0	39	$4.444e-02$	$2.813e-01$
5	4	0	138	$6.978e-01$	$2.813e-01$
5	5	0	122	$6.165e-01$	$2.813e-01$
5	6	0	124	$9.365e-01$	$2.813e-01$
5	7	0	109	$8.554e-01$	$2.813e-01$
5	8	0	106	$7.559e-01$	$2.813e-01$
3	1	2	6	$3.739e-03$	$1.125e+00$
3	2	2	6	$7.507e-03$	$1.125e+00$
3	3	2	6	$2.630e-02$	$1.125e+00$
3	4	2	6	$3.106e-02$	$1.125e+00$
3	5	2	6	$8.293e-02$	$1.125e+00$
3	6	2	6	$1.157e-01$	$1.125e+00$
4	1	2	19	$2.092e-02$	$5.625e-01$
4	2	2	19	$2.632e-02$	$5.625e-01$
4	3	2	19	$4.722e-02$	$5.625e-01$
4	4	2	19	$8.923e-02$	$5.625e-01$
4	5	2	20	$1.998e-01$	$5.625e-01$
4	6	2	20	$2.855e-01$	$5.625e-01$

j	ℓ	$\hat{L}$	#it	$\|\|\mathbf{y}_\Lambda\|_\Gamma\|\|_{\ell_2}$	tol
3	2	1	3	$6.718e-03$	$1.125e+00$
3	3	1	3	$9.356e-03$	$1.125e+00$
3	4	1	3	$1.404e-02$	$1.125e+00$
3	5	1	3	$7.079e-02$	$1.125e+00$
3	6	1	3	$1.104e-01$	$1.125e+00$
3	7	1	3	$1.153e-01$	$1.125e+00$
4	2	1	20	$2.020e-02$	$5.625e-01$
4	3	1	20	$3.527e-02$	$5.625e-01$
4	4	1	20	$5.709e-02$	$5.625e-01$
4	5	1	20	$1.210e-01$	$5.625e-01$
4	6	1	21	$2.140e-01$	$5.625e-01$
4	7	1	21	$2.981e-01$	$5.625e-01$
5	2	1	36	$3.727e-02$	$2.813e-01$
5	3	1	36	$6.812e-02$	$2.813e-01$
5	4	1	36	$1.346e-01$	$2.813e-01$
5	5	1	40	$2.275e-01$	$2.813e-01$
5	6	1	42	$3.323e-01$	$2.813e-01$
5	7	1	90	$3.420e-01$	$2.813e-01$
3	0	3	8	$1.114e-02$	$1.125e+00$
3	1	3	8	$1.391e-02$	$1.125e+00$
3	2	3	8	$1.789e-02$	$1.125e+00$
3	3	3	8	$3.336e-02$	$1.125e+00$
3	4	3	8	$5.736e-02$	$1.125e+00$
3	5	3	8	$1.360e-01$	$1.125e+00$

Table 5.1: Iteration numbers for solving (5.4.26) by CG method, diagonal preconditioning as in (5.6.2) with scaling weights from (5.7.1); #it: number of CG iterations; $\|\mathbf{y}_\Lambda|_\Gamma\|_{\ell_2}$ approximation of trace on Γ.

Γ can be defined via periodization, we can take $\ell_0 = 0$ as the coarsest level of the multiresolution spaces on the boundary. Up to parametric lifting, the single scale bases ${}^2\Phi_\ell$ consist of standard piecewise linear functions so that the approximation order on Γ is 2. The dual multiresolution on Γ is also chosen to be exact of order 2. Likewise one could employ periodic wavelet bases on $\Box$. Here we use tensor products again of the biorthogonal multiresolution spaces of order $d_\Box = 2$ and $\tilde{d}_\Box = 2$ for the dual multiresolution constructed in [DKU2]. Thus, the trial spaces on $\Box$ are standard bilinear finite elements, and the single scale bases are ${}^1\Phi_j$. Due to the adaptation to the boundary of $\Box$, we take $j_0 = 3$ as the lowest level on $\Box$.

Recall from (5.4.3) that for $H_{1,0}, H_{2,0}$ as above, the cases (I), respectively (II), apply. On account of (3.2.67), one can therefore choose the canonical weights

$$d_{1,\lambda} = 2^{|\lambda|}, \quad \lambda \in \mathit{I\!I}_1, \qquad d_{2,\lambda} = 2^{-|\lambda|/2}, \quad \lambda \in \mathit{I\!I}_2, \tag{5.7.1}$$

in the new inner products (5.4.1) and for the corresponding diagonal scaling in (5.6.2). As mentioned previously in Section 3.2.4, an alternative choice is

$$d_{1,\lambda} := A_{1,1}({}^1\psi_\lambda, {}^1\psi_\lambda) = \langle \nabla\, {}^1\psi_\lambda, \nabla\, {}^1\psi_\lambda\rangle_\square + \langle {}^1\psi_\lambda, {}^1\psi_\lambda\rangle_\square, \quad \lambda \in I\!I_1. \tag{5.7.2}$$

For simplicity, we consider only hierarchies of uniformly refined trial spaces on $\square$ and Γ as described in Chapter 3. As pointed out in Section 5.6, it then suffices to compute up to diagonal scaling the sparse blocks

$$\langle \nabla\, {}^1\Phi_{j+\hat{L}}, \nabla\, {}^1\Phi_j\rangle_\square + \langle {}^1\Phi_{j+\hat{L}}, {}^1\Phi_j\rangle_\square, \quad \langle {}^1\Phi_{j+\hat{L}}, {}^2\Phi_\ell\rangle_\Gamma, \quad \langle {}^2\Phi_{\ell+\hat{L}}, {}^1\Phi_j\rangle_\Gamma$$

to obtain then the blocks of the matrices $\mathbf{A}_{\mathbf{J}+\hat{L},\mathbf{J}}$ via (5.6.2). The computation of these matrices is supported by the tools from [BKU].

According to Section 5.6, we use nested iteration based on the conjugate gradient method applied to the corresponding hierarchy of linear systems (5.4.26) where the application of $\mathbf{O}_{\mathbf{J}}^{\mathbf{J}+\hat{L}}$ to a vector has been described at the end of Section 5.6. The iteration on each refinement level stops as soon as the ℓ_2 norm of the residual is smaller than $\mathtt{tol}\|\mathbf{D}_{\mathbf{J}+\hat{L}}^{-1}\mathbf{F}_{\mathbf{J}+\hat{L}}\|_{\ell_2}$. Recall that by (5.6.3) the error of the current iteration in the discrete energy norm induced by $\mathbf{O}_{\mathbf{J}}^{\mathbf{J}+\hat{L}}$ remains uniformly proportional to the residual measured in the ℓ_2 norm. Moreover, also by (5.6.3) the scaling by $\|\mathbf{D}_{\mathbf{J}+\hat{L}}^{-1}\mathbf{F}_{\mathbf{J}+\hat{L}}\|_{\ell_2}$ should reflect the H^1 norm of the solution u. As for the choice of `tol`, recall from Section 5.3 and Theorem 5.1 from [DKS2] that the discretization error is bounded in this case by

$$\inf_{v\in S({}^1\Phi_j)} \|y - v\|_{H^1(\square)} + \inf_{q\in S({}^2\Phi_\ell)} \|p - q\|_{H^{-1/2}(\Gamma)}. \tag{5.7.3}$$

For the current choice of trial functions, the H^1 error for the component y should decay like 2^{-j} while for the current smooth boundary the normal traces remain smooth so that the second component of the error should decay like $2^{-5\ell/2}$. Choosing the error of the first component as target accuracy, we set here `tol`$= \hat{\delta}2^{-j}$ for fixed $\hat{\delta} < 1$, i.e., on each level we have to reduce the error by a factor of at least $1/2$.

Although it would be reasonable to choose ℓ and j so that both terms in (5.7.3) balance, we have applied this scheme for several different choices of j, ℓ, mainly to see the effect on stability. The results are recorded in Table 5.1 for the scaling (5.7.1) and in Table 5.2 for the energy weighting (5.7.2). The first and second columns indicate the refinement levels on $\square$ and Γ. The third column displays the level $\hat{L}$ of expanded truncation. We have taken the same $\hat{L}$ for both components of the solution. The fourth column shows the number of iterations needed to meet the desired accuracy. The fifth column contains the resulting approximation to the trace of the approximate solution on Γ computed as in Section 4.3 while the last column shows the tolerance used to control the iterations. One can see that for $\hat{L} = 0$, which corresponds to symmetric truncation, relatively large iteration numbers are encountered when refining the discretization. Already $\hat{L} = 1$ shows an improvement of stability as predicted, although a somewhat larger $\hat{L}$ is expected to be necessary for full level independent stability. Moreover, recall from Section 4.2.4 that the Galerkin scheme for the saddle point problem (5.3.12) is only stable when the LBB condition is satisfied which, in turn, restricts the mesh size on the boundary relative to that on the domain [DK2]. Here one can see that j and ℓ can be taken rather independently of each other.

One can see that the scaling (5.7.2) gives rise to fewer conjugate gradient iterations. This indicates that the corresponding constants in (2.18), (2.22) have smaller ratios and hence provide tighter bounds for the condition numbers in (5.6.3). This agrees with the observations in [CM2] which refer to the case where homogeneous boundary conditions are incorporated in the trial spaces.

A further series of tests has been undertaken to test the performance of the least squares approach for the saddle point problem in comparison to the technique in Chapter 4. In fact, it has been observed for least squares methods using finite elements that the discrete solutions are often affected by over–diffusion. For the present least squares formulation of problem (4.3.1), we have now chosen the same setup and stopping criterion as for the computations for Table 4.5. The results are displayed in Table 5.3 for $\hat{L} = 0, 1$ and in Table 5.4 for $\hat{L} = 2, 3$. Comparing the numbers for the same level j and ℓ, we see that although the residual $\|\mathbf{r}_\Lambda\|_{\ell_2}$ is of a similar quantitative order, the accuracy of the discrete solution $\mathbf{y}_\Lambda$ restricted to the boundary measured by $\|\mathbf{y}_\Lambda|_\Gamma\|_{\ell_2}$ appears to be at least two orders of magnitude less. For this example, this seems to confirm the experiences made with finite elements.

j	ℓ	$\hat{L}$	#it	$\|\mathbf{y}_\Lambda\|_\Gamma\|_{\ell_2}$	tol
4	3	0	15	$1.996e-02$	$5.625e-01$
4	4	0	15	$5.126e-02$	$5.625e-01$
4	5	0	15	$9.514e-02$	$5.625e-01$
4	6	0	15	$2.364e-01$	$5.625e-01$
4	7	0	15	$3.036e-01$	$5.625e-01$
4	8	0	15	$3.143e-01$	$5.625e-01$
5	3	0	25	$5.549e-02$	$2.813e-01$
5	4	0	25	$8.268e-02$	$2.813e-01$
5	5	0	26	$1.680e-01$	$2.813e-01$
5	6	0	26	$2.905e-01$	$2.813e-01$
5	7	0	26	$3.934e-01$	$2.813e-01$
5	8	0	67	$1.058e+00$	$2.813e-01$
3	1	2	3	$3.024e-03$	$1.125e+00$
3	2	2	3	$1.110e-02$	$1.125e+00$
3	3	2	3	$3.438e-02$	$1.125e+00$
3	4	2	3	$4.385e-02$	$1.125e+00$
3	5	2	3	$1.175e-01$	$1.125e+00$
3	6	2	3	$1.517e-01$	$1.125e+00$
4	1	2	12	$3.043e-02$	$5.625e-01$
4	2	2	12	$3.658e-02$	$5.625e-01$
4	3	2	12	$5.767e-02$	$5.625e-01$
4	4	2	12	$1.112e-01$	$5.625e-01$
4	5	2	12	$2.259e-01$	$5.625e-01$
4	6	2	12	$3.196e-01$	$5.625e-01$

j	ℓ	$\hat{L}$	#it	$\|\mathbf{y}_\Lambda\|_\Gamma\|_{\ell_2}$	tol
3	2	1	2	$8.372e-03$	$1.125e+00$
3	3	1	2	$1.304e-02$	$1.125e+00$
3	4	1	2	$2.109e-02$	$1.125e+00$
3	5	1	2	$8.319e-02$	$1.125e+00$
3	6	1	2	$1.281e-01$	$1.125e+00$
3	7	1	2	$1.337e-01$	$1.125e+00$
4	2	1	13	$2.467e-02$	$5.625e-01$
4	3	1	13	$5.012e-02$	$5.625e-01$
4	4	1	13	$7.234e-02$	$5.625e-01$
4	5	1	13	$1.530e-01$	$5.625e-01$
4	6	1	13	$2.463e-01$	$5.625e-01$
4	7	1	13	$3.182e-01$	$5.625e-01$
5	2	1	26	$5.058e-02$	$2.813e-01$
5	3	1	26	$8.140e-02$	$2.813e-01$
5	4	1	26	$1.569e-01$	$2.813e-01$
5	5	1	26	$2.751e-01$	$2.813e-01$
5	6	1	26	$3.967e-01$	$2.813e-01$
5	7	1	26	$5.161e-01$	$2.813e-01$
3	0	3	2	$4.957e-03$	$1.125e+00$
3	1	3	2	$6.337e-03$	$1.125e+00$
3	2	3	2	$8.417e-03$	$1.125e+00$
3	3	3	2	$1.456e-02$	$1.125e+00$
3	4	3	2	$3.579e-02$	$1.125e+00$
3	5	3	2	$5.857e-02$	$1.125e+00$

Table 5.2: Iteration numbers for solving (5.4.26) by CG method, preconditioning as in (5.6.2) with scaling weights from (5.7.2); #it: number of CG iterations; $\|\mathbf{y}_\Lambda\|_\Gamma\|_{\ell_2}$ approximation of trace on Γ.

j $\square$	ℓ Γ	$\hat{L}$	$\|\|\mathbf{r}_\Lambda\|\|_{\ell_2}$	$\|\|\mathbf{y}_\Lambda\|_\Gamma\|\|_{\ell_2}$
4	3	0	$5.1465e-02$	$1.7312e+00$
4	4	0	$5.9433e-02$	$1.6476e+00$
4	5	0	$3.0936e-02$	$1.5495e+00$
4	6	0	$1.3879e-02$	$9.2745e-01$
4	7	0	$6.0410e-03$	$2.1557e-01$
4	8	0	$3.2308e-03$	$2.0343e-01$
5	3	0	$3.0271e-02$	$1.0565e+00$
5	4	0	$2.8780e-02$	$4.4944e-01$
5	5	0	$2.7503e-02$	$5.3676e-01$
5	6	0	$1.4442e-02$	$4.2148e-01$
5	7	0	$7.6229e-03$	$2.8791e-01$
5	8	0	$3.6950e-03$	$2.6032e-01$
3	2	1	$1.1489e-01$	$1.0707e-02$
3	3	1	$1.1747e-01$	$1.5512e-02$
3	4	1	$5.1633e-02$	$9.1300e-01$
3	5	1	$1.6738e-02$	$1.2754e+00$
3	6	1	$1.5306e-02$	$1.6507e+00$
3	7	1	$6.9586e-03$	$9.5347e-01$
4	2	1	$5.8989e-02$	$7.0244e-01$
4	3	1	$5.8714e-02$	$5.3636e-01$
4	4	1	$5.3311e-02$	$7.0519e-01$
4	5	1	$2.4251e-02$	$4.1096e-01$
4	6	1	$1.3739e-02$	$5.3261e-01$
4	7	1	$7.1939e-03$	$4.7181e-01$
5	2	1	$3.0842e-02$	$4.1659e-01$
5	3	1	$2.9216e-02$	$3.1158e-01$
5	4	1	$2.7908e-02$	$3.9618e-01$

Table 5.3: Iteration numbers for solving (5.4.26) by CG method, diagonal preconditioning as in (5.6.2) with scaling weights from (5.7.1); $\|\mathbf{r}_\Lambda\|_{\ell_2}$ residual; $\|\mathbf{y}_\Lambda|_\Gamma\|_{\ell_2}$ approximation of trace on Γ.

j $\square$	ℓ Γ	$\hat{L}$	$\|\mathbf{r}_\Lambda\|_{\ell_2}$	$\|\mathbf{y}_\Lambda\|_\Gamma\|_{\ell_2}$
3	1	2	$9.9304e-02$	$5.4569e-03$
3	2	2	$1.0128e-01$	$1.8364e-02$
3	3	2	$1.0868e-01$	$5.2847e-02$
3	4	2	$4.9383e-02$	$3.2823e-01$
3	5	2	$2.8174e-02$	$7.8098e-01$
3	6	2	$1.4263e-02$	$4.5308e-01$
4	1	2	$5.4075e-02$	$3.2826e-01$
4	2	2	$5.5078e-02$	$3.5278e-01$
4	3	2	$5.9428e-02$	$3.7308e-01$
4	4	2	$5.4670e-02$	$2.8473e-01$
4	5	2	$3.0017e-02$	$3.8042e-01$
3	0	3	$1.0296e-01$	$2.2825e-02$
3	1	3	$1.0367e-01$	$2.8274e-02$
3	2	3	$1.0380e-01$	$3.5595e-02$
3	3	3	$1.0744e-01$	$6.2837e-02$
3	4	3	$4.6913e-02$	$1.6436e-01$
3	5	3	$1.8635e-02$	$5.7460e-01$

Table 5.4: Iteration numbers for solving (5.4.26) by CG method, preconditioning as in (5.6.2) with scaling weights from (5.7.2); $\|\mathbf{r}_\Lambda\|_{\ell_2}$ residual; $\|\mathbf{y}_\Lambda|_\Gamma\|_{\ell_2}$ approximation of trace on Γ.

6 Control Problems

6.1 Introduction

Finally, the techniques discussed in the previous chapters are applied to control problems involving elliptic boundary value problems. That is, we consider boundary value problems which are affected on (part of) the boundary $\partial\Omega$ by some *control.* This control, in turn, is determined by minimizing some quadratic functional involving the *natural norms* of the *state* of the system and the control. Since these natural norms involve at least one 'broken' norm, they are usually avoided in finite element methods. The optimality conditions of such a minimization problem lead to a representation of the control in terms of the state. Together with the boundary value problem these conditions constitute a linear operator equation consisting of two *weakly coupled* saddle point problems. We will derive this system in Section 6.2 in the context of the general saddle point problems discussed in Section 4.1.1 and show that it is well–posed.

From the discretization of the entire system in terms of (properly scaled) wavelets we infer that it is more convenient to *replace* the original functional by a functional which is more appropriate for computations. The specific form of this functional will emerge from the subsequent discussions. In doing so, we exploit here the fact that it is often more important to capture the typical behavior of a class of functionals than computing with one that is specifically tailored to one format only. The main idea is based on Step 2 of the general concept concentrated in Chapter 2. Since the wavelet concept allows to represent continuous operator equations as ℓ_2–automorphisms, we can consequently formulate the quadratic functional in terms of ℓ_2–norms. This functional is still *equivalent* (in a sense to be detailed below) to the original functional involving the natural norms of the state and the control.

For the numerical solution of the resulting coupled system, the issue of stability of the finite–dimensional discretization will be discussed in Section 6.4 in the context of least squares methods, applying the techniques from Section 5.5. The fact that the least squares methodology transforms the system into a symmetric positive definite system will be further exploited in Section 6.5 which deals with the *iterative solution* of the linear system.

If one wants to avoid solving the coupled optimal control system as a whole, one can resort to techniques solving each of the two saddle point systems alternately. That is, prescribing some state, one computes the control, uses this control to update the state, and so on. Since the boundary control changes frequently, appending boundary conditions by Lagrange multipliers is the method of choice for setting up the system for the state. A convergence proof for a gradient method which is based on solving each of the systems by a *direct* method has been given in [GL1]. There, of course, the emerging fill–in limits the size of the problems. We will discuss here in Section 6.6 concepts for employing wavelet-based iterative solvers instead. For a *fully iterative* scheme *convergence* is established. Moreover, it will be proved that the fully iterative scheme is in combination with a nested iteration strategy an *asymptotically optimal* method in the sense that it provides the solution of the linear system with an amount of iterations *proportional* to the number of unknowns.

In the following we denote by $\Gamma \subseteq \partial\Omega$ that part of the boundary of Ω where a boundary control of Dirichlet type u is effective, maintaining the previously used notation but now allowing u to *vary*. We will consider quadratic functionals which are minimized subject to an elliptic boundary value problem in the Fictitious Domain–Lagrange Multiplier formulation (4.2.13).

There are different points of contact of the work presented in this chapter to previous investigations of optimal control problems. To mention a few, for shape optimization problems involving free boundaries, the combination with Lagrange multiplier techniques and fictitious domain methods is particularly promising, see e.g. [DH, Has, HHK, HasN, KP, NT, Toi]. Least squares techniques for control problems involving, in particular, Navier–Stokes equations have been discussed in [Bo]. In an abstract framework optimal control problems with elliptic systems as constraints are considered and discretized in terms of spectral elements in [AN]. An adaptive Finite Element Galerkin method for quadratic optimal control problems governed by elliptic boundary value problems is analyzed in [BKR]. Optimal control problems with parabolic PDEs as constraints, reducing the problem to Differential–Algebraic Equations by finite–difference and finite element techniques, are discussed in [GJL, PRG]. The question of preconditioning saddle point problems in optimal control has been investigated in [GMPS]. Numerical optimization for control problems with semilinear elliptic equations subject to inequality constraints and discretized by finite differences can be found in [MM]. Newton–SQP methods for control problems with semilinear parabolic equations are discussed in [Tr].

In order to get an idea of the type of problems that can be treated by the techniques proposed here, at this point we turn to a few typical examples.

Example 6.1 On $\Omega = \square$ we consider the elliptic second order boundary value problem (4.2.1),

$$\begin{aligned} -\nabla \cdot (\mathbf{a}\nabla y) + ky &= f && \text{in } \Omega, \\ y &= u && \text{on } \Gamma, \\ (\mathbf{a}\nabla y)\cdot \mathbf{n} &= 0 && \text{on } \Gamma_N, \end{aligned}$$

in its Fictitious Domain–Lagrange Multiplier formulation (4.2.13) from Chapter 4: Find $(y,p) \in H^1(\square) \times (H^{1/2}(\Gamma))'$ such that

$$\begin{aligned} a(y,v) + b(v,p) &= \langle f, v\rangle_\square && \text{for all } v \in H^1(\square), \\ b(y,q) &= \langle u, q\rangle_\Gamma && \text{for all } q \in (H^{1/2}(\Gamma))' \end{aligned}$$

holds. Let the set of admissible controls be chosen as the full space $U_{\text{ad}} = H^{1/2}(\Gamma)$. Assuming that $f \in (H^1(\square))'$ is given, the *optimal control problem* is to find a boundary control $u \in U_{\text{ad}}$ such that (4.2.13) has a unique solution y and in addition the functional

$$\widehat{\mathcal{J}}(y,u) = \tfrac{\omega_1}{2}\,\|y - y_\square\|^2_{H^1(\square)} + \tfrac{\omega_2}{2}\,\|u\|^2_{H^{1/2}(\Gamma)} \tag{6.1.1}$$

is minimized. Here $y_\square$ is some prescribed value of y on $\square$, and y is the solution of (4.2.13) generated by the control input u. Moreover, $0 < \omega_1, \omega_2 < \infty$ are fixed parameters balancing the two norms. ∎

Example 6.2 Another typical example is to consider the same problem but with the functional $\widehat{\mathcal{J}}$ in (6.1.1) replaced by

$$\widehat{\mathcal{J}}(y,u) = \tfrac{\omega_1}{2}\,\|y - y_{\Gamma_y}\|^2_{H^{1/2}(\Gamma_y)} + \tfrac{\omega_2}{2}\,\|u\|^2_{H^{1/2}(\Gamma)}, \tag{6.1.2}$$

where $\Gamma_y \subseteq \partial\Omega$ is some subset of $\partial\Omega$ and $y_{\Gamma_y} \in H^{1/2}(\Gamma_y)$ is given. Such a functional is employed when measurements of y can only be taken on (part of) the boundary. If $\Gamma_y \cap \Gamma \neq \emptyset$ this means that y and thereby also u is fixed on the intersection. Thus, in order to be able to formulate a control problem, we require that $\Gamma \setminus \Gamma_y$ has a strictly positive surface measure. ■

Example 6.3 A more involved example is a transmission problem derived from a coupled *solid/fluid temperature control problem* similar to the one treated in [GL1].

Let the domain $\Omega = \square$ with boundary $\partial\Omega$ consist of two parts, the *solid body domain* Ω_1 and the *fluid flow domain* Ω_2 which are connected by an interface wall Γ_y, see Figure 6.1 for a two-dimensional sketch of the domain and its boundaries. One wants to find a control u on Γ ($\subset \partial\Omega_2$) in order to approximately match a desired temperature of a fluid along (part of) the interface Γ_y. That is, the temperature y is supposed to satisfy the energy equations

$$\begin{aligned} -\hat{\kappa}_1\,\Delta y &= g_1 && \text{in } \Omega_1, \\ -\hat{\kappa}_2\,\Delta y + (\vec{w}\cdot\nabla)y &= g_2(\vec{w}) && \text{in } \Omega_2, \end{aligned} \tag{6.1.3}$$

with boundary conditions

$$\begin{aligned} y &= u && \text{on } \Gamma, \\ \partial_{\mathbf{n}} y &= 0 && \text{on } \Gamma_r, \end{aligned}$$

where $\Gamma_r = \partial\Omega \backslash \Gamma$. The constants $\hat{\kappa}_1, \hat{\kappa}_2$ denote the thermal conductivity coefficients and g_1, g_2 are given functions where $g_2(\vec{w})$ depends nonlinearly on $\vec{w}$. Here $\vec{w} : \Omega_2 \to I\!R^n$ is the fluid velocity field on Ω_2 which is determined beforehand (assuming that the viscous fluid is incompressible) by the Stokes equations on Ω_2. That is, $\vec{w} : \Omega_2 \to I\!R^n$ and the pressure $p : \Omega_2 \to I\!R$ are related by

$$\begin{aligned} -\nu\,\Delta\vec{w} + \nabla p &= \vec{f} && \text{in } \Omega_2, \\ \operatorname{div}\vec{w} &= 0 && \text{in } \Omega_2, \end{aligned} \tag{6.1.4}$$

with boundary conditions

$$\begin{aligned} && \vec{w} &= \vec{w}_0 && \text{on } \Gamma, \\ \vec{w} = \mathbf{0} \quad \text{on } \Gamma_y \cup \Gamma_b, && \mathbf{n}\cdot\nabla\vec{w} &= \mathbf{0} && \text{on } \Gamma_o, \end{aligned} \tag{6.1.5}$$

where Γ and Γ_o are the control inflow and the outflow boundary, respectively. Furthermore, $\vec{w}_0$ is given and the constant ν is the kinematic viscosity coefficient of the fluid, and Γ_b is the bottom of Ω_2. For the numerical solution of the Stokes equations (6.1.4)

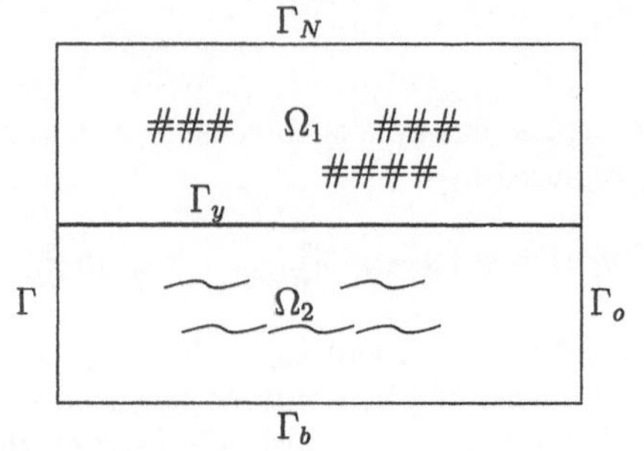

Figure 6.1: sketch of Ω in two dimensions

with (6.1.5) one can employ one's favorite finite difference or finite element code, or the Galerkin method using biorthogonal wavelets proposed in [DKU1].

Then one is left with solving the equations for the temperature (6.1.3) in Ω, where on Ω_2 the velocity field $\vec{w}$ enters the differential equations as variable but known coefficient and in the source term on the right hand side.

In the framework of the previously used weak formulation (4.2.13), the governing equations on Ω_1 and Ω_2 can be set up in weak formulation as before. We dispense at this point with a detailed derivation and refer to [KK]. The cost functional to be minimized is of the form

$$\widehat{\mathcal{J}}(y,u) = \tfrac{\omega_1}{2} \|y - y_{\Gamma_y}\|^2_{H^{1/2}(\Gamma_y)} + \tfrac{\omega_2}{2} \|u\|^2_{H^{1/2}(\Gamma)} \tag{6.1.6}$$

where y_{Γ_y} is a prescribed value on Γ_y.

In [GL1] the following case is treated for spatial dimension $n = 2$: the boundary control is assumed to be in $\{w \in H^1(\Gamma) : w = 0 \text{ on } \overline{\Gamma} \cap \overline{\partial\Omega_1}\}$ and the functional $\widehat{\mathcal{J}}$ contains the norms $\|\cdot\|_{H^1(\Gamma)}$ and $\|\cdot\|_{L_2(\Gamma_y)}$ for the control u and the state y, respectively. ∎

Here we will formulate the control problem using the *natural energy functional*, measuring y and u in their 'natural' norms. These norms are identified for y to be either $\|\cdot\|_{H^1(\Box)}$ or $\|\cdot\|_{H^{1/2}(\Gamma_y)}$ and for the control u the norm is $\|\cdot\|_{H^{1/2}(\Gamma)}$. As in the least squares case, the wavelet framework is in contrast to standard approaches a convenient way for the evaluation of the norms for the 'negative' and 'broken' spaces.

In the next section, we recall some theoretical results on control problems covering the above examples. To this end, we formulate the elliptic boundary value problem again as a saddle point problem in abstract form as in Section 4.1.1 and recall the optimality conditions for the minimization of a quadratic functional. Their explicit form leads to the requirement to solve an additional *adjoint system* also in form of a saddle point problem. It is then shown that this coupled system fits into the general concept from Chapter 2, namely, that the resulting operator provides an isomorphism between suitable function spaces. Using wavelet bases for the relevant function spaces, Step 2 of the general concept will be applied to derive a discrete ℓ_2 system. We will see then that formulating the coupled system as a least squares problem yields stability of the discrete system so that no compatibility conditions like the LBB condition are necessary.

For the sake of completeness, it should be mentioned that also systems with *distributed control*, i.e., control in the right hand side like

$$\begin{aligned} a(y,v)+b(v,p) &= \langle u,v\rangle_{\Box} && \text{for all } v\in H^1(\Box),\\ b(y,q) &= \langle g,q\rangle_{\Gamma} && \text{for all } q\in (H^{1/2}(\Gamma))', \end{aligned} \tag{6.1.7}$$

can be treated by the methods proposed here. In fact, in this case the techniques in the next section lead to an adjoint system which plays the role of the primal system here, and vice versa, see [Li, Hal]. In this sense the problems with boundary or distributed control are equivalent.

6.2 The Continuous Case: Two Coupled Saddle Point Problems

In this section, we embed problems of the above type into the abstract setting for saddle point problems from Section 4.1.1 in terms of Hilbert spaces Y and Q, and operators A, B. Assuming that in addition to some prescribed $f \in Y'$ the function u is given, Theorem 4.1 provides existence and uniqueness of the solution of the abstract saddle point problem (4.1.9): Given $(f,u) \in Y' \times Q'$, find $(y,p) \in Y \times Q$ such that

$$\mathcal{L}\begin{pmatrix} y\\ p\end{pmatrix} \equiv \begin{pmatrix} A & B'\\ B & 0\end{pmatrix}\begin{pmatrix} y\\ p\end{pmatrix} = \begin{pmatrix} f\\ u\end{pmatrix}. \tag{6.2.1}$$

Up to this point, the function u has been assumed to be *known*. We now view the system (4.1.9) from the point of control theory.

In order to treat the different examples considered above in a unified way, in addition to Y and Q, a third Hilbert space $H(Y)$ with inner product $(\cdot,\cdot)_{H(Y)}$ and induced norm $\|\cdot\|_{H(Y)}$ is introduced. The norm $\|\cdot\|_{H(Y)}$ is the energy norm in which the state y is measured or *observed*, hence, $H(Y)$ is called the *observation space* for y. Thus, naturally the minimization functional would be formulated in terms of $\|\cdot\|_{H(Y)}$ and $\|\cdot\|_{Q'}$. The full space $U_{\text{ad}} = Q'$ is the natural set of *admissible* controls. In order to formulate the functional, it will be assumed that there is a linear continuous operator

$$Z : Y \to H(Y).$$

In view of the above examples where $Y = H^1(\Box)$ and $Q = (H^{1/2}(\Gamma))'$, we distinguish two choices of $H(Y)$ and Z.

Remark 6.4 *(i) If $H(Y) = Y$, then Z is just the identity. The functional (6.1.1) falls into this category.*

(ii) When measurements controlling the state can only be taken on (part of) the boundary $\partial\Omega$ denoted by Γ_y, $H(Y)$ is chosen as the trace space,

$$H(Y) = H^{1/2}(\Gamma_y),$$

and

$$Z : H^1(\Box) \to H^{1/2}(\Gamma_y)$$

is the standard trace operator onto Γ_y. *Recall then that for this situation also the Trace Theorem, Corollary 4.10, applies: for any* $f \in H^1(\Box)$, *one can estimate*

$$\|Zf\|_{H^{1/2}(\Gamma_y)} \leq c_{T_1,y} \|f\|_{H^1(\Box)}. \tag{6.2.2}$$

Conversely, for every $h \in H^{1/2}(\Gamma_y)$, *there exists some* $f \in H^1(\Box)$ *such that* $Zf = h$ *and*

$$\|f\|_{H^1(\Box)} \leq c_{T_2,y}\|h\|_{H^{1/2}(\Gamma_y)}. \tag{6.2.3}$$

This case covers the functionals (6.1.2), (6.1.6).

A natural definition of a minimization functional would now involve the norms $\|\cdot\|_{H(Y)}$ and $\|\cdot\|_{Q'}$. However, in many situations one is interested in the *qualitative* behavior of a system rather than in the values in particular norms so that it suffices to replace these norms by suitable *equivalent* norms

$$\|\cdot\|_1^2 := (\cdot,\cdot)_1 \sim \|\cdot\|_{H(Y)}^2, \qquad \|\cdot\|_2^2 := (\cdot,\cdot)_2 \sim \|\cdot\|_{Q'}^2, \tag{6.2.4}$$

which are numerically easier to evaluate. In the context of least–squares methods we have encountered already norms of the format (5.4.1) which are evaluated in terms of wavelets. Now we define the *minimization functional* as

$$\mathcal{J}(y,u) := \frac{\omega}{2}\|Zy - y_\Box\|_1^2 + \frac{1}{2}\|u\|_2^2 \tag{6.2.5}$$

involving some constant weight $0 < \omega < \infty$ and some prescribed value $y_\Box$. We explicitly write the operator Z to keep track of it.

The general form of the *minimization problem* considered here is then the following:

find $(y,p,u) \in Y \times Q \times Q'$ such that $\mathcal{J}(y,u)$ defined in (6.2.5) is minimized subject to (6.2.1).	(6.2.6)

In order to explore existence and uniqueness of a solution of (6.2.6), set

$$\begin{aligned} \mathcal{G}(y,p,u) &:= \mathcal{L}\begin{pmatrix} y \\ p \end{pmatrix} - \begin{pmatrix} f \\ u \end{pmatrix} = \left(\mathcal{L} \,\middle|\, \begin{matrix} 0 \\ -I \end{matrix} \right) \begin{pmatrix} y \\ p \\ u \end{pmatrix} - \begin{pmatrix} f \\ 0 \end{pmatrix}, \\ \mathcal{K} &:= \{(y,p,u) \in Y \times Q \times Q' : \ \mathcal{G}(y,p,u) = 0\}. \end{aligned} \tag{6.2.7}$$

Denote by $D^s\hat{\mathcal{J}}(u; v_1, \ldots, v_s)$ the s-th variation of $\hat{\mathcal{J}}$ at u in directions $v_1, \ldots, v_s$, see e.g. [Z]. In particular, for $s = 1$

$$D\hat{\mathcal{J}}(u;v) = \langle \delta\hat{\mathcal{J}}(u), v\rangle := \lim_{t\to 0} \frac{\hat{\mathcal{J}}(u+tv) - \hat{\mathcal{J}}(u)}{t} \tag{6.2.8}$$

is the (Gateaux) derivative of $\hat{\mathcal{J}}$ at u in direction v.

Remark 6.5 *Let A, B be such that the conditions of Theorem 4.1 hold. Then $\mathcal{G}$ is a* submersion, *see [Z], Definition 43.15. In fact, $X := Y \times Q \times Q'$ is a product of Hilbert spaces and $\mathcal{G}$ is continuously differentiable on all of X. Furthermore, $\mathcal{G}' : X \to Y' \times Q'$ is surjective, i.e.,* $\text{range}(\mathcal{G}'(y,p,u)) = Y' \times Q'$. *This follows from Theorem 4.1 since $\mathcal{L}$ is invertible.*

The following generalized Weierstrass theorem is a special case of Theorem 43.D from [Z]. It gives necessary and sufficient conditions in terms of derivatives of the *Lagrangian functional*

$$\begin{aligned}\textsc{Lagr}(y,p,u,z,\mu) \; &:= \; \tfrac{\omega}{2}\|Zy - y_\square\|_1^2 + \tfrac{1}{2}\|u\|_2^2 \\ &\quad + \left\langle (z,\mu), \begin{pmatrix} A & B' \\ B & 0 \end{pmatrix} \begin{pmatrix} y \\ p \end{pmatrix} - \begin{pmatrix} f \\ u \end{pmatrix} \right\rangle \end{aligned} \qquad (6.2.9)$$

which is defined on $Y \times Q \times Q' \times Y \times Q$. It is formed by appending the conditions (6.2.1) by means of additional Lagrange multipliers z, μ to the minimization functional (6.2.5). But first it is useful to rewrite $\mathcal{J}$ as a functional of the control u alone. This means representing y in terms of u. Observe that the first equation in (6.2.1) can be written as

$$y = -A^{-1}B'p + A^{-1}f, \qquad (6.2.10)$$

which yields by insertion into the second equation in (6.2.1)

$$BA^{-1}B'p = BA^{-1}f - u. \qquad (6.2.11)$$

Substituting p into (6.2.10) yields

$$\begin{aligned} y = y(u) \; &= \; A^{-1}B'(BA^{-1}B')^{-1}u + A^{-1}\left(I - B'(BA^{-1}B')^{-1}BA^{-1}\right)f \\ &=: \; A^{-1}(\widetilde{B}'u + \widetilde{f}). \end{aligned} \qquad (6.2.12)$$

Observe that $\widetilde{B} = ((B(A^{-1})'B')^{-1}B$ inherits the surjectivity property of B,

$$\ker \widetilde{B}' = \{0\} \qquad (6.2.13)$$

since B is surjective and $BA^{-1}B'$ is invertible. Inserting this into (6.2.5) gives

$$\mathcal{J}(u) = \mathcal{J}(y(u),u) \; = \; \tfrac{\omega}{2}\|ZA^{-1}(\widetilde{B}'u + \widetilde{f}) - y_\square\|_1^2 + \tfrac{1}{2}\|u\|_2^2. \qquad (6.2.14)$$

Theorem 6.6 *Suppose that the following two conditions hold:*

(i) $\mathcal{J} : X \to \mathbb{R}$ is differentiable at $x^* \in X$, and X is a real Banach space;

(ii) $\mathcal{G}$ is a submersion at x^*.

Then, if $\mathcal{J}$ has a (free) local minimum at x^* with respect to $\mathcal{K}$, then Lagr *satisfies the necessary conditions*

$$\delta\textsc{Lagr}(x^*) = 0. \qquad (6.2.15)$$

If $\mathcal{J}$ is strongly convex, then (6.2.15) is also sufficient for the unique minimum of (6.2.6) to be attained at x^*.

Recall from [Z] that the functional $\mathcal{J}$ is said to have a *free* local minimum at x^* with respect to $\mathcal{K}$ if $\mathcal{J}(x^*) \leq \mathcal{J}(x)$ for all $x \in U(x^*) \cap \mathcal{K}$ where $U(x^*)$ is a full neighborhood of x^* in X.

Remark 6.7 *The functional $\mathcal{J}$ in (6.2.14) is twice differentiable on Q' with derivatives*

$$\begin{aligned} D\mathcal{J}(u;v) &= \omega\,(Zy(u) - y_\Box,\, ZA^{-1}\widetilde{B}'v)_1 + (u,v)_2 \\ &= \omega\,(ZA^{-1}(\widetilde{B}'u + \widetilde{f}) - y_\Box,\, ZA^{-1}\widetilde{B}'v)_1 + (u,v)_2 \end{aligned} \qquad (6.2.16)$$

for all $v \in Q'$ and

$$D^2\mathcal{J}(u;v,w) = \omega\,(ZA^{-1}\widetilde{B}'v,\, ZA^{-1}\widetilde{B}'w)_1 + (v,w)_2 \qquad (6.2.17)$$

for all $v, w \in Q'$. In particular, $\mathcal{J}$ is strictly convex on Q', i.e.,

$$D^2\mathcal{J}(u;v,v) > 0 \qquad \textit{for any } v \in Q' \setminus \{0\}. \qquad (6.2.18)$$

Proof: For any $u, v \in Q'$ and $t > 0$, one has for $\mathcal{J}$ from (6.2.14)

$$\begin{aligned} \frac{\mathcal{J}(u+tv) - \mathcal{J}(u)}{t} &= \tfrac{\omega}{2}\left(2(Zy(u) - y_\Box,\, ZA^{-1}\widetilde{B}'v)_1 + t\,\|ZA^{-1}\widetilde{B}'v\|_2^2\right) \\ &\quad + \tfrac{1}{2}\,(2(u,v)_2 + t\|v\|_2^2) \end{aligned}$$

yielding

$$D\mathcal{J}(u;v) = \lim_{t\to 0} \frac{\mathcal{J}(u+tv) - \mathcal{J}(u)}{t} = \omega(Zy(u) - y_\Box,\, ZA^{-1}\widetilde{B}'v)_1 + (u,v)_2$$

and thus (6.2.16) upon inserting (6.2.12). Furthermore, let $\hat{\mathcal{G}}(u;v) := \langle \delta\mathcal{J}(u), v\rangle$. Then

$$\frac{\hat{\mathcal{G}}(u+tw;v) - \hat{\mathcal{G}}(u;v)}{t} = \omega\,(ZA^{-1}\widetilde{B}'v,\, ZA^{-1}\widetilde{B}'w)_2 + (v,w)_2$$

is independent of u, yielding the identity (6.2.17) for all $v, w \in Q'$. In particular, one has

$$D^2\mathcal{J}(u;v,v) = \omega\|ZA^{-1}\widetilde{B}'v\|_1^2 + \|v\|_2^2 \geq \|v\|_2^2 > 0 \qquad (6.2.19)$$

for all nonzero $v \in Q'$. ∎

In order to apply Theorem 6.6, it will be convenient to employ the Riesz maps $\mathcal{R}_1 : H(Y) \to H(Y)'$, $\mathcal{R}_2 : Q' \to Q$ defined by

$$\begin{aligned} \langle \mathcal{R}_1 v, w\rangle &:= (v,w)_1 \qquad &&\text{for all } v, w \in H(Y), \\ \langle \mathcal{R}_2 v, w\rangle &:= (v,w)_2 \qquad &&\text{for all } v, w \in Q', \end{aligned} \qquad (6.2.20)$$

see (5.2.11). Note that in view of (6.2.4) one has

$$\|\mathcal{R}_1 v\|_{H(Y)'} \sim \|v\|_{H(Y)}, \qquad \|\mathcal{R}_2 v\|_Q \sim \|v\|_{Q'}. \qquad (6.2.21)$$

Now the Lagrangian functional (6.2.9) can be rewritten in the form

$$\begin{aligned}\textsc{Lagr}(y,p,u,z,\mu) &= \tfrac{\omega}{2}\langle\mathcal{R}_1(Zy-y_\Box), Zy-y_\Box\rangle + \tfrac{1}{2}\langle\mathcal{R}_2u,u\rangle \\ &+ \left\langle (z,\mu), \begin{pmatrix} A & B' \\ B & 0 \end{pmatrix}\begin{pmatrix} y \\ p\end{pmatrix} - \begin{pmatrix} f \\ u \end{pmatrix}\right\rangle \end{aligned} \tag{6.2.22}$$

In view of Remark 6.5, we will consider only those cases of operators where A, B are such that $\mathcal{G}$ is a submersion on all of X.

Lemma 6.8 *Let $\mathcal{J}$ be the functional in (6.2.14), and let $f \in Y'$ and $y_\Box$ be given. Then the Euler equations for the minimization problem (6.2.6) are equivalent to*

$$\begin{aligned} \mathcal{N}U &:= \begin{pmatrix} \mathcal{L} & \mathcal{E} \\ \hat{\mathcal{E}} & \hat{\mathcal{L}} \end{pmatrix}\begin{pmatrix} y \\ p \\ z \\ u \end{pmatrix} \\ &:= \left(\begin{array}{cc|cc} A & B' & 0 & 0 \\ B & 0 & 0 & -I \\ \hline \omega Z'\mathcal{R}_1 Z & 0 & A' & (\mathcal{R}_2B)' \\ 0 & 0 & \mathcal{R}_2B & 0 \end{array}\right)\begin{pmatrix} y \\ p \\ z \\ u \end{pmatrix} \\ &= \begin{pmatrix} f \\ 0 \\ \omega Z'\mathcal{R}_1 y_\Box \\ 0 \end{pmatrix} =: F. \end{aligned} \tag{6.2.23}$$

Proof: According to Theorem 6.6, we need to determine the Euler equations (6.2.15),

$$\delta\textsc{Lagr}(y,p,u,z,\mu;r) = 0 \qquad \text{for } r = y,p,u,z,\mu. \tag{6.2.24}$$

This yields for $r = y, p, u, z, \mu$

$$\begin{aligned} \omega Z'\mathcal{R}_1 Zy - \omega Z'\mathcal{R}_1 y_\Box + A'z + B'\mu &= 0, \\ Bz &= 0, \\ \mathcal{R}_2 u - \mu &= 0, \\ Ay + B'p - f &= 0, \\ By - u &= 0. \end{aligned} \tag{6.2.25}$$

One can further eliminate from the third equation the variable μ,

$$\mu = \mathcal{R}_2 u,$$

and insert it into the first equation to yield

$$\begin{aligned} A'z + (\mathcal{R}_2B)'u &= -\omega Z'\mathcal{R}_1(Zy - y_\Box) \\ Bz &= 0, \\ Ay + B'p &= f, \\ By &= u. \end{aligned} \tag{6.2.26}$$

Multiplying the second equation by $\mathcal{R}_2$ which does not change the solution z since $\mathcal{R}_2$ is an isomorphism gives the system (6.2.23). This latter step is done to generate a lower right block in $\mathcal{N}$ which is symmetric. ∎

The variable z appearing in (6.2.23) is often called the *adjoint state variable.*

Since $\mathcal{J}$ is quadratic, it will be sufficient to solve the *Euler equations* (6.2.15) in order to compute the unique solution of (6.2.6).

A few further remarks are in order.

Remark 6.9 *(i) Note that (6.2.23) can be interpreted as solving simultaneously two systems of saddle point problems, namely, the* primal system *(6.2.1)*

$$\begin{pmatrix} A & B' \\ B & 0 \end{pmatrix} \begin{pmatrix} y \\ p \end{pmatrix} = \begin{pmatrix} f \\ u \end{pmatrix}$$

together with the adjoint system

$$\hat{\mathcal{L}} \begin{pmatrix} z \\ u \end{pmatrix} = \begin{pmatrix} A' & (\mathcal{R}_2 B)' \\ \mathcal{R}_2 B & 0 \end{pmatrix} \begin{pmatrix} z \\ u \end{pmatrix} = \begin{pmatrix} -\omega Z' \mathcal{R}_1 (Zy - y_\square) \\ 0 \end{pmatrix} \tag{6.2.27}$$

for $(z, u) \in Y \times Q'$.

(ii) By applying block Gaussian elimination to the adjoint system, one could get an explicit expression for the control u in terms of y which involves the Schur complement and the inverse of A. Of course, the formulation (6.2.23) has the advantage that one does not need to compute a sufficiently accurate approximation of A^{-1} or the inverse of the Schur complement.

(iii) For control problems with distributed control, *the role of the primal and adjoint system is reversed. In this sense, boundary and distributed control problems are equivalent, and the subsequent investigation applies to both variants.*

From Theorem 4.1 and the fact that $\mathcal{R}_1$, $\mathcal{R}_2$ are Riesz operators, we can immediately infer that the system (6.2.27) has a unique solution.

Corollary 6.10 *If the inf–sup condition (4.1.10) for A and the inf–sup condition for B (4.1.11) hold, there exists a unique solution of (6.2.27), that is,*

$$\hat{\mathcal{L}} : Y \times Q' \to Y' \times Q, \tag{6.2.28}$$

is an isomorphism, and one has the equivalence

$$\left\| \begin{pmatrix} v \\ q \end{pmatrix} \right\|_{Y \times Q'} \sim \left\| \hat{\mathcal{L}} \begin{pmatrix} v \\ q \end{pmatrix} \right\|_{Y' \times Q} \tag{6.2.29}$$

for any $(v, q) \in Y \times Q'$.

By Theorem 4.1 and Corollary 6.10, we can derive corresponding estimates for the weakly coupled system (6.2.23).

Corollary 6.11 *Let the inf–sup assumption (4.1.10) for A and the inf–sup condition (4.1.11) for B hold. Then $\mathcal{N}$ is an isomorphism*

$$\mathcal{N} : \mathcal{H} := Y \times Q \times Y \times Q' \longrightarrow \mathcal{H}', \tag{6.2.30}$$

with the equivalence

$$\|V\|_{\mathcal{H}} \sim \|\mathcal{N}V\|_{\mathcal{H}'} \qquad \textit{for any } V \in \mathcal{H}. \tag{6.2.31}$$

Proof: The invertibility of $\mathcal{N}$ follows already from Theorem 6.6 and Remark 6.7 since $\mathcal{G}$ is, under the conditions on A and B, a submersion. For deriving the estimate (6.2.31), one has for any $V = (y, p, z, u)^T \in \mathcal{H}$ by definition of the norms (2.4)

$$\begin{aligned} \|\mathcal{N}V\|^2_{\mathcal{H}'} &= \left\| \begin{pmatrix} \mathcal{L}\binom{y}{p} + \mathcal{E}\binom{z}{u} \\ \hat{\mathcal{E}}\binom{y}{p} + \hat{\mathcal{L}}\binom{z}{u} \end{pmatrix} \right\|^2_{\mathcal{H}'} = \left\| \begin{pmatrix} \mathcal{L}\binom{y}{p} + \binom{0}{-u} \\ \binom{\omega Z'\mathcal{R}_1 Zy}{0} + \hat{\mathcal{L}}\binom{z}{u} \end{pmatrix} \right\|^2_{\mathcal{H}'} \\ &\lesssim \left\| \mathcal{L}\begin{pmatrix} y \\ p \end{pmatrix} \right\|^2_{Y' \times Q'} + \|u\|^2_{Q'} + \left\| \hat{\mathcal{L}}\begin{pmatrix} z \\ u \end{pmatrix} \right\|^2_{Y' \times Q} + \omega^2 \|Z'\mathcal{R}_1 Zy\|^2_{Y'}. \end{aligned} \tag{6.2.32}$$

If $H(Y) = Y$, then $Z = I$, and (6.2.21) yields that the last term on the right hand side can be estimated by

$$\omega \|Z'\mathcal{R}_1 Zy\|_{Y'} \lesssim \omega \|y\|_Y. \tag{6.2.33}$$

In the second case discussed in Remark 6.4(ii), one can estimate this term using the trace estimates (6.2.2), (6.2.3) and (6.2.21),

$$\omega \|Z'\mathcal{R}_1 Zy\|_{Y'} \lesssim \omega c_{T_2,y} c_{T_1,y} \|y\|_Y. \tag{6.2.34}$$

Combining (6.2.33) or (6.2.34) with (6.2.32) yields

$$\|\mathcal{N}V\|^2_{\mathcal{H}'} \lesssim \left\| \mathcal{L}\begin{pmatrix} y \\ p \end{pmatrix} \right\|^2_{Y' \times Q'} + \|u\|^2_{Q'} + \left\| \hat{\mathcal{L}}\begin{pmatrix} z \\ u \end{pmatrix} \right\|^2_{Y' \times Q} + \omega^2 \|y\|^2_Y. \tag{6.2.35}$$

By (4.1.13) and (6.2.29) one therefore obtains

$$\|\mathcal{N}V\|^2_{\mathcal{H}'} \lesssim \|V\|^2_{\mathcal{H}} + \|u\|^2_{Q'} + \omega^2 \|y\|^2_Y \lesssim \|V\|^2_{\mathcal{H}}. \tag{6.2.36}$$

Since $\mathcal{N}$ is linear, continuous and invertible on a product of Hilbert spaces, it follows from the inverse mapping theorem that $\mathcal{N}^{-1}$ is also linear and continuous, i.e., the estimate

$$\|V\|_{\mathcal{H}} \lesssim \|\mathcal{N}V\|_{\mathcal{H}'}$$

holds. ∎

In summary, Step 1 of the general concept from Chapter 2 is established for the weakly coupled system of saddle point problems (6.2.23).

Remark 6.12 *One could also think of formulating the optimization problem (6.2.6) as a least squares problem as follows. Recalling the definition of $\mathcal{J}$ in (6.2.5), consider the least squares functional*

$$\widehat{LS}(v,q,u) := \mathcal{J}(v,q) + \left\| \mathcal{L}\begin{pmatrix} v \\ q \end{pmatrix} - \begin{pmatrix} f \\ u \end{pmatrix} \right\|_{Y' \times Q'}$$

which is on account of relations corresponding to (6.2.4) for $\|\cdot\|_{Y'}, \|\cdot\|_{Q'}$ equivalent to

$$LS(V) := LS(v,q,u) = \mathcal{J}(v,q) + \|Av + B'q - f\|_3^2 + \|Bv - u\|_4^2.$$

However, this approach does not *produce the same solution of the optimization problem (6.2.6) since it only requires the constraints (6.2.1) to be satisfied in a least squares sense. One could enforce these by introducing weights as penalty parameters which are increased accordingly.*

Instead a different least squares formulation will be used to solve the discretized and preconditioned versions of (6.2.23).

6.3 Discretization and Preconditioning

We have asserted already that biorthogonal wavelet bases with the properties listed in Section 3 exist for the spaces Y, Q. Thus, we can discretize and precondition (6.2.23) in the sense of Step 2, similarly as in Section 4.1.2. In fact, the primal system (6.2.1) will exactly attain the form (4.1.20) when expanding (y,p) in terms of the weighted wavelet bases (4.1.19),

$$(y,p)^T = (\mathbf{y}^T \mathbf{D}_Y^{-1} \Psi_Y, \, \mathbf{p}^T \mathbf{D}_Q^{-1} \Psi_Q)^T,$$

yielding

$$\mathbf{L}\begin{pmatrix} \mathbf{y} \\ \mathbf{p} \end{pmatrix} \equiv \begin{pmatrix} \mathbf{A} & \mathbf{B}^T \\ \mathbf{B} & \mathbf{0} \end{pmatrix} \begin{pmatrix} \mathbf{y} \\ \mathbf{p} \end{pmatrix} = \begin{pmatrix} \mathbf{f} \\ \mathbf{u} \end{pmatrix}$$

with the abbreviations from (4.1.21),

$$\begin{aligned} \mathbf{A} &= \mathbf{D}_Y^{-1} \langle \Psi_Y, A\Psi_Y \rangle \mathbf{D}_Y^{-1}, & \mathbf{f} &= \mathbf{D}_Y^{-1} \langle \Psi_Y, f \rangle, \\ \mathbf{B} &= \mathbf{D}_Q^{-1} \langle \Psi_Q, B\Psi_Y \rangle \mathbf{D}_Y^{-1}, & \mathbf{u} &= \mathbf{D}_Q^{-1} \langle \Psi_Q, u \rangle. \end{aligned}$$

However, in the setup of the dual system (6.2.27) discretizations of the Riesz operators $\mathcal{R}_1$ and $\mathcal{R}_2$ would come into play. Thus, working in wavelet coordinates suggests like in Section 4.2.4 to start out by *formulating* the optimization problem *already in terms of the discrete ℓ_2 norms* to extract the main features of the approach. This means that instead of the functional (6.2.5) we select the quadratic *minimization functional* as

$$\mathbf{J}(\mathbf{y}, \mathbf{u}) := \tfrac{\omega}{2} \|\mathbf{Z}\mathbf{y} - \mathbf{y}_\square\|_{\ell_2}^2 + \tfrac{1}{2} \|\mathbf{u}\|_{\ell_2}^2, \tag{6.3.1}$$

where $\mathbf{y}_\square$ descends from expanding $y_\square$ in terms of $\Psi_{H(Y)}$, $y_\square = \mathbf{y}_\square^T \mathbf{D}_{H(Y)}^{-1} \Psi_{H(Y)}$, and $\mathbf{Z}$ is the discrete ℓ_2–mapping representing the (properly normalized) operator Z,

$$\mathbf{Z} := \mathbf{D}_{H(Y)}^{-1} \langle \Psi_{H(Y)}, Z\Psi_Y \rangle \mathbf{D}_Y^{-1}. \tag{6.3.2}$$

Thus, the most convenient form of the minimization problem that still captures the main features of the functional, namely, to measure effects in natural norms, but that allows to drop unimportant terms like mass matrices can be formulated in the following *discrete form:*

$$\boxed{\text{Find } (\mathbf{y},\mathbf{u}) \in \ell_2(I\!I_Y \times I\!I_Q) \text{ such that } \mathbf{J}(\mathbf{y},\mathbf{u}) \text{ defined in (6.3.1) is minimized subject to (4.1.20).}} \tag{6.3.3}$$

Remark 6.13 *The minimization problems (6.2.6) and (6.3.3) or, precisely, the quadratic functionals (6.2.5) and (6.3.1) are* equivalent *in the sense that the involved norms are equivalent, i.e., with the selected scaling (4.1.19) we have*

$$\|v\|_1 \sim \|\mathbf{v}\|_{\ell_2}, \qquad \|q\|_2 \sim \|\mathbf{q}\|_{\ell_2}. \tag{6.3.4}$$

On account of (6.2.4), we infer that (6.3.3) is also equivalent to a minimization problem involving a quadratic functional with norms $\|\cdot\|_{H(Y)}$ *and* $\|\cdot\|_{Q'}$.

However, we stress that, because of the shift into another Sobolev space, problem (6.3.3) is not *equivalent to minimize a functional of the form*

$$\tfrac{\omega}{2}\|Zy - y_\square\|^2_{L_2} + \tfrac{1}{2}\|u\|^2_{L_2}$$

involving L_2 norms, subject to the constraints (4.1.20).

Consequently, in the following we will exclusively treat the new minimization problem (6.3.3).

The previous investigations, Lemma 6.8 yield that the optimization problem (6.3.3) is solved by solving in addition to (4.1.20) the system

$$\mathbf{L}^T\begin{pmatrix}\mathbf{z}\\ \mathbf{u}\end{pmatrix} \equiv \begin{pmatrix}\mathbf{A}^T & \mathbf{B}^T\\ \mathbf{B} & 0\end{pmatrix}\begin{pmatrix}\mathbf{z}\\ \mathbf{u}\end{pmatrix} = -\omega\begin{pmatrix}\mathbf{Z}^T(\mathbf{Z}\mathbf{y}-\mathbf{y}_\square)\\ 0\end{pmatrix} \tag{6.3.5}$$

where z has been expanded as $z = \mathbf{z}^T\mathbf{D}_Y^{-1}\Psi_Y$.

Remark 6.14 *Note that the linear operator appearing in (6.3.5) is indeed the* adjoint *of the operator* $\mathbf{L}$ *in (4.1.20). A discretization of (6.2.27) instead would have become*

$$\begin{pmatrix}\mathbf{A} & (\mathbf{R}_2\mathbf{B})^T\\ \mathbf{R}_2\mathbf{B} & 0\end{pmatrix}\begin{pmatrix}\mathbf{z}\\ \mathbf{u}\end{pmatrix} = -\omega\begin{pmatrix}\mathbf{Z}^T\mathbf{R}_1(\mathbf{Z}\mathbf{y}-\mathbf{y}_\square)\\ 0\end{pmatrix},$$

where $\mathbf{R}_1, \mathbf{R}_2$ are the discretized Riesz operators. Their explicit form could be determined as in (5.4.15) in terms of (scaled) mass matrices $\langle\Psi_{H(Y)},\Psi_{H(Y)}\rangle$ and $\langle\tilde{\Psi}_Q,\tilde{\Psi}_Q\rangle$.

As in the derivation of the system (6.2.23) for the problem (6.2.6), the Euler equations for the new problem (6.3.3) are determined by applying the same techniques to deduce the assertions in Theorem 6.6, Remark 6.7 and Lemma 6.8. Since the following discrete version of the assertions in Remark 6.7 will be employed later also in the analysis of fully iterative methods for the discrete systems, we derive their precise form here.

First we rewrite the functional $\mathbf{J}$ again as a function of $\mathbf{u}$ alone. Representing $\mathbf{y}$ in terms of $\mathbf{u}$ as in (6.2.12),

$$\begin{aligned}\mathbf{y} = \mathbf{y}(\mathbf{u}) &= \mathbf{A}^{-1}\mathbf{B}^T(\mathbf{B}\mathbf{A}^{-1}\mathbf{B}^T)^{-1}\mathbf{u} + \mathbf{A}^{-1}\left(\mathbf{I} - \mathbf{B}^T(\mathbf{B}\mathbf{A}^{-1}\mathbf{B}^T)^{-1}\mathbf{B}\mathbf{A}^{-1}\right)\mathbf{f} \\ &=: \mathbf{A}^{-1}(\widetilde{\mathbf{B}}^T\mathbf{u} + \widetilde{\mathbf{f}}) \end{aligned} \tag{6.3.6}$$

and inserting this into (6.3.1) gives

$$\begin{aligned}\mathbf{J}(\mathbf{u}) = \mathbf{J}(\mathbf{y}(\mathbf{u}),\mathbf{u}) &= \tfrac{\omega}{2}\|\mathbf{Z}\mathbf{y}(\mathbf{u}) - \mathbf{y}_\square\|_{\ell_2}^2 + \tfrac{1}{2}\|\mathbf{u}\|_{\ell_2}^2 \\ &= \tfrac{\omega}{2}\|\mathbf{Z}\mathbf{A}^{-1}(\widetilde{\mathbf{B}}^T\mathbf{u} + \widetilde{\mathbf{f}}) \ - \mathbf{y}_\square\|_{\ell_2}^2 + \tfrac{1}{2}\|\mathbf{u}\|_{\ell_2}^2. \end{aligned} \tag{6.3.7}$$

Let again $D^s\mathbf{J}(\mathbf{u};\mathbf{v}_1,\ldots,\mathbf{v}_s)$ denote the s-th variation of $\mathbf{J}$ at $\mathbf{u}$ in directions $\mathbf{v}_1,\ldots,\mathbf{v}_s$, where in particular

$$D\mathbf{J}(\mathbf{u};\mathbf{v}) = \langle \delta\mathbf{J}(\mathbf{u}),\mathbf{v}\rangle = \lim_{t\to 0}\frac{\mathbf{J}(\mathbf{u}+t\mathbf{v}) - \mathbf{J}(\mathbf{u})}{t}. \tag{6.3.8}$$

The discrete version of Remark 6.7 now reads as follows.

Remark 6.15 *The functional* $\mathbf{J}$ *defined in (6.3.7) is twice differentiable on* ℓ_2 *with derivative*

$$\begin{aligned} D\mathbf{J}(\mathbf{u};\mathbf{v}) &= \omega\langle \mathbf{Z}\mathbf{y}(\mathbf{u}) - \mathbf{y}_\square,\ \mathbf{Z}\mathbf{A}^{-1}\widetilde{\mathbf{B}}^T\mathbf{v}\rangle + \langle\mathbf{u},\mathbf{v}\rangle \\ &= \omega\langle \mathbf{Z}\mathbf{A}^{-1}(\widetilde{\mathbf{B}}^T\mathbf{u} + \widetilde{\mathbf{f}}) - \mathbf{y}_\square,\ \mathbf{Z}\mathbf{A}^{-1}\widetilde{\mathbf{B}}^T\mathbf{v}\rangle + \langle\mathbf{u},\mathbf{v}\rangle \end{aligned} \tag{6.3.9}$$

for all $\mathbf{v}\in\ell_2$. *We infer explicitly*

$$\delta\mathbf{J}(\mathbf{u}) = \omega\widetilde{\mathbf{B}}\mathbf{A}^{-T}\mathbf{Z}^T(\mathbf{Z}\mathbf{y}(\mathbf{u}) - \mathbf{y}_\square) + \mathbf{u}. \tag{6.3.10}$$

Moreover, the second derivative is

$$D^2\mathbf{J}(\mathbf{u};\mathbf{v},\mathbf{w}) = \omega\langle \mathbf{Z}\mathbf{A}^{-1}\widetilde{\mathbf{B}}^T\mathbf{v},\ \mathbf{Z}\mathbf{A}^{-1}\widetilde{\mathbf{B}}^T\mathbf{w}\rangle + \langle\mathbf{v},\mathbf{w}\rangle \tag{6.3.11}$$

for all $\mathbf{v},\mathbf{w}\in\ell_2$. *In particular,* $\mathbf{J}$ *is strictly convex on* ℓ_2, *i.e.,*

$$D^2\mathbf{J}(\mathbf{u};\mathbf{v},\mathbf{v}) > 0 \qquad \text{for any } \mathbf{v}\in\ell_2\setminus\{\mathbf{0}\}. \tag{6.3.12}$$

Proof: The assertions (6.3.9), (6.3.11) and (6.3.12) follow exactly as in the proof of Remark 6.7. The expression for $\delta\mathbf{J}(\mathbf{u})$ is derived from the representation (6.3.9) and the definition (6.3.8),

$$\begin{aligned}\langle\delta\mathbf{J}(\mathbf{u}),\mathbf{v}\rangle &= \omega\langle \mathbf{Z}\mathbf{y}(\mathbf{u}) - \mathbf{y}_\square,\ \mathbf{Z}\mathbf{A}^{-1}\widetilde{\mathbf{B}}^T\mathbf{v}\rangle + \langle\mathbf{u},\mathbf{v}\rangle \\ &= \omega\langle \widetilde{\mathbf{B}}\mathbf{A}^{-T}\mathbf{Z}^T(\mathbf{Z}\mathbf{y}(\mathbf{u}) - \mathbf{y}_\square),\mathbf{v}\rangle + \langle\mathbf{u},\mathbf{v}\rangle\end{aligned} \tag{6.3.13}$$

where $\mathbf{y}$ is determined in (6.3.6). ∎

The analog to Corollary 6.11 covered by Step 2 of the general concept applied to the two saddle point systems (4.1.20) and (6.3.5) is then the following result.

Corollary 6.16 *The operator* $\mathbf{N}$ *defined by*

$$\mathbf{N}\mathbf{U} := \begin{pmatrix} \mathbf{L} & \mathbf{E} \\ \hat{\mathbf{E}} & \mathbf{L}^T \end{pmatrix} \begin{pmatrix} \mathbf{y} \\ \mathbf{p} \\ \mathbf{z} \\ \mathbf{u} \end{pmatrix} := \left(\begin{array}{cc|cc} \mathbf{A} & \mathbf{B}^T & \mathbf{0} & \mathbf{0} \\ \mathbf{B} & \mathbf{0} & \mathbf{0} & -\mathbf{I} \\ \hline \omega \mathbf{Z}^T\mathbf{Z} & \mathbf{0} & \mathbf{A}^T & \mathbf{B}^T \\ \mathbf{0} & \mathbf{0} & \mathbf{B} & \mathbf{0} \end{array}\right) \begin{pmatrix} \mathbf{y} \\ \mathbf{p} \\ \mathbf{z} \\ \mathbf{u} \end{pmatrix} \tag{6.3.14}$$

$$= \begin{pmatrix} \mathbf{f} \\ \mathbf{0} \\ -\omega\, \mathbf{Z}^T \mathbf{y}_{\square} \\ \mathbf{0} \end{pmatrix} =: \mathbf{F}$$

is an ℓ_2*–automorphism,*

$$\ell_2 = \ell_2(I\!I) := \ell_2(I\!I_Y \times I\!I_{Q'} \times I\!I_Y \times I\!I_{Q'}),$$

i.e., for any $\mathbf{V} \in \ell_2$ *the equivalence*

$$\|\mathbf{N}\mathbf{V}\|_{\ell_2} \sim \|\mathbf{V}\|_{\ell_2} \tag{6.3.15}$$

holds.

To ensure stability of the finite–dimensional discrete systems, the truncation techniques from Chapter 5 will be applied. To this end, we first rewrite the weakly coupled infinite–dimensional system (6.3.14) as a least squares problem as follows. For $\mathbf{N}$ playing the role of $\mathbf{A}$ in (5.4.13), Corollary 6.16 yields Theorem 5.6 for the system (6.3.14).

Corollary 6.17 *The vector* $\mathbf{U}$ *solves (6.3.14) if and only if* $\mathbf{U}$ *solves the system*

$$\mathbf{P}\mathbf{U} := \mathbf{N}^T\mathbf{N}\mathbf{U} = \mathbf{N}^T\mathbf{F}. \tag{6.3.16}$$

Moreover, the matrix $\mathbf{P}$ *defines an automorphism of* $\ell_2 = \ell_2(I\!I)$ *and one has*

$$\mathbf{V}^T\mathbf{P}\mathbf{V} = \|\mathbf{N}\mathbf{V}\|^2_{\ell_2} \sim \|\mathbf{V}\|^2_{\ell_2}. \tag{6.3.17}$$

6.4 The Discrete Finite–Dimensional Problem

We are now precisely in the situation of Section 5.4 where the original saddle point problem has been transformed into an infinite–dimensional system which is well-posed in Euclidean metric $\ell_2(I\!I)$. Recalling the notation of Chapter 5 we now have $M = 4$, and we will apply the truncation technique introduced in Section 5.5.

To this end, consider again any finite set

$$\boldsymbol{\Lambda} = \Lambda_1 \times \cdots \times \Lambda_4 = \Lambda_Y \times \Lambda_Q \times \Lambda_Y \times \Lambda_{Q'} \subset I\!I \tag{6.4.1}$$

with elements $\boldsymbol{\lambda} = (\lambda_1, \ldots, \lambda_4)$, and let $\hat{\boldsymbol{\Lambda}} \subset I\!I$ be another such finite set.

Recalling that the truncated blocks of $\mathbf{N}$ relative to $\mathbf{\Lambda}, \hat{\mathbf{\Lambda}}$ are defined as in (5.4.24), the computable version of the weakly coupled saddle point problem (6.3.14) is the following:

Discrete System of Weakly Coupled Saddle Point Problems WSPP($\hat{\mathbf{\Lambda}}, \mathbf{\Lambda}$): *Given finite sets $\hat{\mathbf{\Lambda}}, \mathbf{\Lambda} \subset \mathbb{I}$, set*

$$\mathbf{P}_{\mathbf{\Lambda}}^{\hat{\mathbf{\Lambda}}} := \mathbf{N}_{\hat{\mathbf{\Lambda}},\mathbf{\Lambda}}^T \mathbf{N}_{\hat{\mathbf{\Lambda}},\mathbf{\Lambda}} \tag{6.4.2}$$

and find $\mathbf{U}_{\mathbf{\Lambda}} \in \mathbb{R}^{\mathbf{\Lambda}}$ such that

$$\mathbf{P}_{\mathbf{\Lambda}}^{\hat{\mathbf{\Lambda}}} \mathbf{U}_{\mathbf{\Lambda}} = \mathbf{N}_{\hat{\mathbf{\Lambda}},\mathbf{\Lambda}}^T \mathbf{F}_{\hat{\mathbf{\Lambda}}}, \tag{6.4.3}$$

where

$$\mathbf{F}_{\mathbf{\Lambda}} := \mathbf{D}_{\mathbf{\Lambda}}^{-1} \begin{pmatrix} \langle \Psi_{\Lambda_1}^1, f \rangle \\ \mathbf{0} \\ -\omega \langle \Psi_{\Lambda_3}^3, Z' y_\square \rangle \\ \mathbf{0} \end{pmatrix} =: \mathbf{D}_{\mathbf{\Lambda}}^{-1} \langle \mathbf{\Psi}_{\mathbf{\Lambda}}, F \rangle, \tag{6.4.4}$$

and $\mathbf{\Psi}_{\mathbf{\Lambda}}$, $\mathbf{U}_{\mathbf{\Lambda}}$ and $\mathbf{D}_{\mathbf{\Lambda}}$ are according to (5.4.28).

We have seen already in Section 5.5 that the estimates (5.5.19) apply for the operators $\mathbf{A}$, $\mathbf{B}$ appearing in $\mathbf{L}$. In exactly the same way, this is shown for $\mathbf{Z}$. In fact, in the first case corresponding to Remark 6.4(i), $\mathbf{Z} = \mathbf{I}$. In the second case where $\mathbf{Z}$ represents the regular trace operator onto $H^{1/2}(\Gamma_y)$, one is again in the situation (5.5.15). Thus, since all blocks $\mathbf{A}^{i,l}$ appearing in (6.3.14) satisfy (5.5.19), Proposition 5.15 applies for the problem WSPP($\hat{\mathbf{\Lambda}}, \mathbf{\Lambda}$). Here $P(\cdot,\cdot)$ and $P^{\hat{\mathbf{\Lambda}}}(\cdot,\cdot)$ are the bilinear forms induced by $\mathbf{P}$ defined according to Section 5.

Corollary 6.18 *For any fixed $\alpha \in (0,1)$ and any $\mathbf{\Lambda}$ in (6.4.1) one can choose some $\hat{\mathbf{\Lambda}}$ such that*

$$\alpha P(V,V) \le P^{\hat{\mathbf{\Lambda}}}(V,V) \le P(V,V), \quad V \in \mathcal{S}_{\mathbf{\Lambda}}, \tag{6.4.5}$$

and

$$\#\hat{\mathbf{\Lambda}} \le c_\# \#\mathbf{\Lambda},$$

where the constant $c_\#$ depends on the constants $c, C, c_{\mathcal{L}}, C_{\mathcal{L}}, \alpha$ but not on $\mathbf{\Lambda}$ and V.

6.5 Iterative Methods for WSPP($\hat{\mathbf{\Lambda}}, \mathbf{\Lambda}$)

After ensuring stability of the discrete system (6.4.3) it remains to discuss strategies for its numerical solution.

Corollary 6.18 yields together with Theorem 5.16 the following fact also for the system WSPP($\hat{\mathbf{\Lambda}}, \mathbf{\Lambda}$).

Corollary 6.19 *Given $\mathbf{\Lambda}$, one can find $\hat{\mathbf{\Lambda}}$ satisfying*

$$\#\hat{\mathbf{\Lambda}} \sim \#\mathbf{\Lambda}$$

such that WSPP$(\hat{\mathbf{\Lambda}}, \mathbf{\Lambda})$ *is stable and the matrices* $\mathbf{P}_{\mathbf{\Lambda}}^{\hat{\mathbf{\Lambda}}}$ *in (6.4.3) are positive definite and have uniformly bounded spectral condition numbers*

$$\kappa(\mathbf{P}_{\mathbf{\Lambda}}^{\hat{\mathbf{\Lambda}}}) \;\lesssim\; \alpha^{-1}. \tag{6.5.1}$$

where α *is the constant from (6.4.5).*

Since $\mathbf{P}_{\mathbf{\Lambda}}^{\hat{\mathbf{\Lambda}}}$ is also symmetric, one can apply any of the algorithms like conjugate gradient method for such types of operators. Furthermore, all results from Section 5.6 apply. In particular, a nested iteration strategy combined with a conjugate gradient method for (6.4.3) yields a method where the solution is obtained by an amount of operations which is proportional to the number of unknowns. In this sense, the algorithm is *asymptotically optimal.*

6.6 Alternative Iterative Methods for the Coupled System

The least squares approach discussed in Sections 6.4 and 6.5 provides Step 3, stability of the discretizations, which is theoretically fully justified for the problem (6.3.14), and leads to a symmetric positive definite problem. Nevertheless, although the constants in (6.5.1) do not depend on the discretizations, they are still squared, see (5.6.3). In addition, the approach seems to amplify a deterioration of the accuracy of the discretization. Thus, one looks for alternative iterative methods that directly involve discrete *finite-dimensional* systems of the form (6.3.14) derived from (6.3.3) and that are *computationally efficient.*

Recall that we have assured stability of the discretizations by satisfying the LBB condition (4.2.26) for finite discretizations of the primal system (6.2.1) by applying the results from Section 4.2.4. In view of Corollary 6.11 this immediately carries over to stability for the whole *finite-dimensional* system indexed for convenience by Λ

$$\mathbf{N}_\Lambda \mathbf{U}_\Lambda \equiv \begin{pmatrix} \mathbf{L}_\Lambda & \mathbf{E}_\Lambda \\ \hat{\mathbf{E}}_\Lambda & \mathbf{L}_\Lambda^T \end{pmatrix} \begin{pmatrix} \mathbf{y}_\Lambda \\ \mathbf{p}_\Lambda \\ \mathbf{z}_\Lambda \\ \mathbf{u}_\Lambda \end{pmatrix} = \left(\begin{array}{cc|cc} \mathbf{A}_\Lambda & \mathbf{B}_\Lambda^T & 0 & 0 \\ \mathbf{B}_\Lambda & 0 & 0 & -\mathbf{I} \\ \hline \omega \mathbf{Z}_\Lambda^T \mathbf{Z}_\Lambda & 0 & \mathbf{A}_\Lambda^T & \mathbf{B}_\Lambda^T \\ 0 & 0 & \mathbf{B}_\Lambda & 0 \end{array}\right) \begin{pmatrix} \mathbf{y}_\Lambda \\ \mathbf{p}_\Lambda \\ \mathbf{z}_\Lambda \\ \mathbf{u}_\Lambda \end{pmatrix} \tag{6.6.1}$$

$$= \begin{pmatrix} \mathbf{f}_\Lambda \\ 0 \\ -\omega\, \mathbf{Z}_\Lambda^T (\mathbf{y}_\Box)_\Lambda \\ 0 \end{pmatrix} \equiv \mathbf{F}_\Lambda.$$

The starting point will be to recall that this finite-dimensional system (6.6.1) is the same as

$$\mathbf{L}_\Lambda \begin{pmatrix} \mathbf{y}_\Lambda \\ \mathbf{p}_\Lambda \end{pmatrix} = \begin{pmatrix} \mathbf{A}_\Lambda & \mathbf{B}_\Lambda^T \\ \mathbf{B}_\Lambda & 0 \end{pmatrix} \begin{pmatrix} \mathbf{y}_\Lambda \\ \mathbf{p}_\Lambda \end{pmatrix} = \begin{pmatrix} \mathbf{f}_\Lambda \\ \mathbf{u}_\Lambda \end{pmatrix} \tag{6.6.2}$$

$$\mathbf{L}_\Lambda^T \begin{pmatrix} \mathbf{z}_\Lambda \\ \mathbf{u}_\Lambda \end{pmatrix} = \begin{pmatrix} \mathbf{A}_\Lambda^T & \mathbf{B}_\Lambda^T \\ \mathbf{B}_\Lambda & 0 \end{pmatrix} \begin{pmatrix} \mathbf{z}_\Lambda \\ \mathbf{u}_\Lambda \end{pmatrix} = -\,\omega \begin{pmatrix} \mathbf{Z}_\Lambda^T(\mathbf{Z}_\Lambda \mathbf{y}_\Lambda - (\mathbf{y}_\Box)_\Lambda) \\ 0 \end{pmatrix}. \tag{6.6.3}$$

The linear system matrix in (6.6.1) is typically very large, indefinite and also unsymmetric since in general $\omega \mathbf{Z}_\Lambda^T \mathbf{Z}_\Lambda \neq -\mathbf{I}$. Recall that all block operators are *sparse* in the sense that they can be applied with an amount of operations which is *linear* in the number of unknowns by realizing the application in terms of the Fast Wavelet Transform, see e.g. (3.2.75). In contrast to the single systems (6.6.2) and (6.6.3), the operator $\mathbf{N}_\Lambda$ does *not* represent a saddle point operator. Nevertheless, because of its particular block structure, there are a few methods that suggest themselves.

6.6.1 Semi–Iterative Methods

The first method, a semi–iterative strategy, has been used in [GL1] for a coupled solid/fluid temperature control problem similar to the one discussed in [KK]. It is based on solving alternately the two systems (6.6.2), (6.6.3) by QR decomposition of $\mathbf{L}_\Lambda$. Once a factorization of $\mathbf{L}_\Lambda$ is available, the solution of both (6.6.2) and (6.6.3) only requires back substitution. It will be illustrative to recall this method here since it will serve as a point of departure for the methods considered later.

ALGORITHM DIRECT

STEP 1: Choose $\mathbf{u}_\Lambda^{(0)}, \mathbf{y}_\Lambda^{(0)}, \mathbf{p}_\Lambda^{(0)}, \mathbf{z}_\Lambda^{(0)}, \boldsymbol{\mu}_\Lambda^{(0)}$; set up the system matrix $\mathbf{L}_\Lambda$ in (6.6.2) and right hand sides in (6.6.2), (6.6.3); set $i = 0$; compute the QR decomposition of $\mathbf{L}_\Lambda$;

STEP 2: set $\mathbf{u}_\Lambda = \mathbf{u}_\Lambda^{(i)}$ and solve (6.6.2) to obtain a new $\mathbf{y}_\Lambda^{(i+1)}$;

STEP 3: update the right hand side of (6.6.3) by setting $\mathbf{y}_\Lambda = \mathbf{y}_\Lambda^{(i+1)}$ and solve

$$\mathbf{L}_\Lambda^T \begin{pmatrix} \mathbf{z}_\Lambda^{(i+1)} \\ \boldsymbol{\mu}_\Lambda^{(i+1)} \end{pmatrix} = -\omega \begin{pmatrix} \mathbf{Z}_\Lambda^T(\mathbf{Z}_\Lambda \mathbf{y}_\Lambda^{(i+1)} - (\mathbf{y}_\square)_\Lambda) \\ \mathbf{0} \end{pmatrix};$$

STEP 4: update $\mathbf{u}_\Lambda^{(i)}$ by computing

$$\mathbf{u}_\Lambda^{(i+1)} := \mathbf{u}_\Lambda^{(i)} - \rho_i \mathbf{K}_\Lambda \mathbf{z}_\Lambda^{(i+1)}, \tag{6.6.4}$$

where ρ_i is some step size parameter determined later and $\mathbf{K}_\Lambda$ is a linear operator approximating the normal derivative of y on Γ;

STEP 5: set $i = i + 1$ and repeat Steps 2, 3, 4 until a prescribed tolerance for $\mathbf{u}_\Lambda$ is reached.

Here the particular form of STEP 4 results from measuring u in $\mathcal{J}$ in $H^1(\Gamma)$. The convergence of ALGORITHM DIRECT is proved in [GL1] for ρ_i in a certain range depending on the second variation of $\mathcal{J}$.

Here we consider a basic algorithm for the general case of controlling u in $\mathcal{J}$ in a norm equivalent to its natural norm (6.2.5). This will serve later as the basis for a fully iterative method. This method is also based on a simple gradient algorithm in terms of the functional $\mathbf{J}$ depending only on $\mathbf{u}_\Lambda$ derived in (6.3.7).

ALGORITHM BASIC

STEP 1: Fix $\mathbf{u}_\Lambda^{(0)}$;

STEP 2: update $\mathbf{u}_\Lambda^{(i)}$ by computing

$$\mathbf{u}_\Lambda^{(i+1)} := \mathbf{u}_\Lambda^{(i)} - \rho_i \,\delta\mathbf{J}(\mathbf{u}_\Lambda^{(i)}) \tag{6.6.5}$$

where ρ_i is some step size parameter determined later;

STEP 3: set $i = i+1$ and repeat Steps 2 until a prescribed tolerance for $\mathbf{u}_\Lambda$ is reached.

The range for the step size parameter ρ_i is detailed below.

For the *evaluation* of $\delta\mathbf{J}(\mathbf{u}_\Lambda^{(i)})$, consider the systems

$$\mathbf{L}_\Lambda \begin{pmatrix} \mathbf{y}_\Lambda^{(i+1)} \\ \mathbf{p}_\Lambda^{(i+1)} \end{pmatrix} = \begin{pmatrix} \mathbf{f}_\Lambda \\ \mathbf{u}_\Lambda^{(i)} \end{pmatrix} \tag{6.6.6}$$

and

$$\mathbf{L}_\Lambda^T \begin{pmatrix} \mathbf{z}_\Lambda^{(i+1)} \\ \boldsymbol{\mu}_\Lambda^{(i+1)} \end{pmatrix} = -\,\omega \begin{pmatrix} \mathbf{Z}_\Lambda^T(\mathbf{Z}_\Lambda \mathbf{y}_\Lambda^{(i+1)} - (\mathbf{y}_\Box)_\Lambda) \\ \mathbf{0} \end{pmatrix}. \tag{6.6.7}$$

As we will see later in Proposition 6.24, for $\mathbf{J}$ from (6.3.7), $\delta\mathbf{J}(\mathbf{u}_\Lambda^{(i)})$ satisfies

$$\delta\mathbf{J}(\mathbf{u}_\Lambda^{(i)}) = \mathbf{u}_\Lambda^{(i)} - \boldsymbol{\mu}_\Lambda^{(i+1)} \tag{6.6.8}$$

where $\boldsymbol{\mu}_\Lambda^{(i+1)}$ solves (6.6.7). Thus, as in ALGORITHM DIRECT each step (6.6.5) requires the successive solution of two saddle point problems which at this point resorts to direct solvers. Proposition 6.23 below assures that ALGORITHM BASIC converges. Of course, solving (6.6.2), (6.6.3) by factoring $\mathbf{L}_\Lambda$ by e.g. QR decomposition severely constrains the range of applications to relatively small problems. Thus, one looks for fully iterative methods to be able to handle also very large problems. There are different strategies one can think of.

6.6.2 Block Kaczmarz Iteration

The following method denoted as *block Kaczmarz iteration* is known to converge for any regular system matrix [Ka]. This method is equivalent to solving

$$\mathbf{N}_\Lambda \mathbf{N}_\Lambda^T \hat{\mathbf{V}}_\Lambda = \mathbf{F}_\Lambda \tag{6.6.9}$$

for $\hat{\mathbf{V}}_\Lambda$ using Gauss-Seidel iterations, and then setting $\mathbf{V}_\Lambda = \mathbf{N}_\Lambda^T \hat{\mathbf{V}}_\Lambda$. Kaczmarz iteration can also be applied to overdetermined linear systems such as the ones arising in tomography. There this method is called ART for 'algebraic reconstruction technique' [Na]. For the interpretation of Kaczmarz' iteration as a Gauss-Seidel method, see e.g. [Ha2]. There are also different block versions that have been investigated in the context of solving overdetermined systems, see e.g. [E, Ni, Ta].

In the present context, these ideas for determining the solution of (6.6.9) can be used in form of a block Gauss-Seidel method applied alternately to the blocks in (6.6.1) which is described as follows. Denote by

$$\mathbf{N}_\Lambda = \begin{pmatrix} (\mathbf{N}_\Lambda)_1 \\ (\mathbf{N}_\Lambda)_2 \end{pmatrix} \tag{6.6.10}$$

the block decomposition of the system matrix in (6.6.1) into two block rows. Let for any rectangular matrix X the matrix X^+ be the *generalized inverse* of X which is uniquely defined by the Moore–Penrose Axioms. Then Kaczmarz iteration consists of alternately projecting the system (6.6.1) onto different hyperplanes generated by the first and second block row in (6.6.1).

ALGORITHM KACZMARZ

STEP 1: Choose $\mathbf{y}_\Lambda^{(0)}, \mathbf{p}_\Lambda^{(0)}, \mathbf{u}_\Lambda^{(0)}, \mathbf{z}_\Lambda^{(0)}$; set up the system matrix $\mathbf{L}_\Lambda$ in (6.6.2) and right hand sides in (6.6.2), (6.6.3); set $i = 0$;

STEP 2: compute the new vector

$$\begin{pmatrix} \mathbf{y}_\Lambda^{(i+1)} \\ \mathbf{p}_\Lambda^{(i+1)} \end{pmatrix} = \begin{pmatrix} \mathbf{y}_\Lambda^{(i)} \\ \mathbf{p}_\Lambda^{(i)} \end{pmatrix} + (\mathbf{N}_\Lambda)_1^+ \left(\begin{pmatrix} \mathbf{f}_\Lambda \\ \mathbf{u}_\Lambda^{(i)} \end{pmatrix} - (\mathbf{N}_\Lambda)_1 \begin{pmatrix} \mathbf{y}_\Lambda^{(i)} \\ \mathbf{p}_\Lambda^{(i)} \end{pmatrix} \right);$$

STEP 3: compute the new vector

$$\begin{pmatrix} \mathbf{z}_\Lambda^{(i+1)} \\ \mathbf{u}_\Lambda^{(i+1)} \end{pmatrix} = \begin{pmatrix} \mathbf{z}_\Lambda^{(i)} \\ \mathbf{u}_\Lambda^{(i)} \end{pmatrix} + (\mathbf{N}_\Lambda)_2^+ \left(-\omega \begin{pmatrix} \mathbf{Z}_\Lambda^T(\mathbf{Z}_\Lambda \mathbf{y}_\Lambda^{(i+1)} - (\mathbf{y}_\Box)_\Lambda) \\ \mathbf{0} \end{pmatrix} - (\mathbf{N}_\Lambda)_2 \begin{pmatrix} \mathbf{z}_\Lambda^{(i)} \\ \mathbf{u}_\Lambda^{(i)} \end{pmatrix} \right);$$

STEP 4: set $i = i + 1$ and repeat Step 2 and 3 until a prescribed tolerance for $\mathbf{y}_\Lambda$ is reached.

In Step 2 and Step 3, one needs to determine the Pseudo–Inverses of the block rows $(\mathbf{N}_\Lambda)_1^+$, $(\mathbf{N}_\Lambda)_2^+$ for which one could employ yet another iterative method. Of course, one would then have to investigate the effect of the approximation of the generalized inverses on the convergence of ALGORITHM KACZMARZ. Note that in view of (6.6.9) the application of $(\mathbf{N}_\Lambda)_1^+$, $(\mathbf{N}_\Lambda)_2^+$ provide a Least– Squares approach which one could continue to analyze.

6.6.3 A Fully Iterative Method for Coupled Saddle Point Problems

An alternative to the previous method is motivated on the one hand by ALGORITHM BASIC and on the other hand by the fact that by now many efficient iterative methods for saddle point problems exist. The aim of this section is to propose a corresponding iteration that will replace the direct solution of the saddle point problems (6.6.2) and (6.6.3) by a standard iterative method for saddle point problems and prove the convergence of the resulting fully iterative method. Based on the convergence results for ALGORITHM BASIC, one would think that such an iteration converges when both systems are solved iteratively within suitable dynamically chosen tolerances.

We formulate the main algorithm first without specifying the type of iterative method which is used for the solution of the saddle point problems (6.6.2) and (6.6.3). We only assume that it converges and call it 'inner iteration', abbreviated as ALGORITHM INNIT. Since the algorithm involves the iterative solution of both systems, it is termed 'outer iteration'.

ALGORITHM OUTIT

STEP 1: Choose $\mathbf{u}_\Lambda^{(0)}, \widetilde{\mathbf{y}}_\Lambda^{(0)}, \widetilde{\mathbf{p}}_\Lambda^{(0)}, \widetilde{\mathbf{z}}_\Lambda^{(0)}, \widetilde{\boldsymbol{\mu}}_\Lambda^{(0)}$, set up the blocks $\mathbf{L}_\Lambda$ in (6.6.2), $\hat{\mathbf{E}}_\Lambda$ in (6.6.3), and $\mathbf{f}_\Lambda$, $(\mathbf{y}_\Box)_\Lambda$; let $\varepsilon_y(i)$, $\varepsilon_\mu(i)$ be some suitable stage–dependent tolerances to be determined later; set $i = 0$;

STEP 2: apply ALGORITHM INNIT to compute an approximate solution $(\widetilde{\mathbf{y}}_\Lambda^{(i+1)}, \widetilde{\mathbf{p}}_\Lambda^{(i+1)})^T$ of (6.6.6) with right hand side $(\mathbf{f}_\Lambda, \mathbf{u}_\Lambda^{(i)})^T$ and initial guesses $(\widetilde{\mathbf{y}}_\Lambda^{(i)}, \widetilde{\mathbf{p}}_\Lambda^{(i)})^T$ that satisfies

$$\left\| \mathbf{L}_\Lambda \begin{pmatrix} \widetilde{\mathbf{y}}_\Lambda^{(i+1)} \\ \widetilde{\mathbf{p}}_\Lambda^{(i+1)} \end{pmatrix} - \begin{pmatrix} \mathbf{f}_\Lambda \\ \mathbf{u}_\Lambda^{(i)} \end{pmatrix} \right\|_{\ell_2} < \varepsilon_y(i+1); \tag{6.6.11}$$

STEP 3: apply ALGORITHM INNIT to compute a solution $(\widetilde{\mathbf{z}}_\Lambda^{(i+1)}, \widetilde{\boldsymbol{\mu}}_\Lambda^{(i+1)})^T$ of (6.6.3) with right hand side $-\omega(\mathbf{Z}_\Lambda^T(\mathbf{Z}_\Lambda \widetilde{\mathbf{y}}_\Lambda^{(i+1)} - (\mathbf{y}_\Box)_\Lambda), \mathbf{0})^T$ and initial guesses $(\widetilde{\mathbf{z}}_\Lambda^{(i)}, \widetilde{\boldsymbol{\mu}}_\Lambda^{(i)})^T$ until

$$\left\| \mathbf{L}_\Lambda^T \begin{pmatrix} \widetilde{\mathbf{z}}_\Lambda^{(i+1)} \\ \widetilde{\boldsymbol{\mu}}_\Lambda^{(i+1)} \end{pmatrix} + \omega \begin{pmatrix} \mathbf{Z}_\Lambda^T(\mathbf{Z}_\Lambda \widetilde{\mathbf{y}}_\Lambda^{(i+1)} - (\mathbf{y}_\Box)_\Lambda) \\ \mathbf{0} \end{pmatrix} \right\|_{\ell_2} < \varepsilon_\mu(i+1); \tag{6.6.12}$$

STEP 4: update $\mathbf{u}_\Lambda^{(i)}$ by

$$\mathbf{u}_\Lambda^{(i+1)} = \mathbf{u}_\Lambda^{(i)} - \rho_i\,(\mathbf{u}_\Lambda^{(i)} - \widetilde{\boldsymbol{\mu}}_\Lambda^{(i+1)}); \tag{6.6.13}$$

STEP 5: set $i = i+1$ and repeat Step 2, 3, 4 until prescribed tolerance for $\mathbf{u}_\Lambda$ is reached.

The choice of the tolerances $\varepsilon_y(i+1), \varepsilon_\mu(i+1)$ as well as of the step size parameter ρ_i will be discussed below. The convergence of ALGORITHM OUTIT is not clear beforehand since the iterative solution of (6.6.2) and (6.6.3) produces an additional error that appears again in the right hand side of the corresponding adjoint system. Thus, in the convergence analysis one needs to assure that the errors produced in the inner iterations do not accumulate and can be fully controlled.

Before we prove convergence of ALGORITHM OUTIT, we prove that ALGORITHM BASIC converges. For convenience, we omit Λ indexing the finite-dimensional quantities until its role has to be taken into account again. But first we state the following observation will be useful later.

Remark 6.20 *In view of the relation (6.3.6), the term $D^2\mathbf{J}$ determined in (6.3.9) can be rewritten as*

$$D^2\mathbf{J}(\mathbf{u};\mathbf{v},\mathbf{w}) \;=\; \omega\,\langle \mathbf{Z}\tilde{\mathbf{y}},\mathbf{Z}\bar{\mathbf{y}}\rangle + \langle \mathbf{v},\mathbf{w}\rangle, \tag{6.6.14}$$

where $\tilde{\mathbf{y}}$ and $\bar{\mathbf{y}}$ solve

$$\mathbf{L}\begin{pmatrix}\tilde{\mathbf{y}}\\ \tilde{\mathbf{p}}\end{pmatrix} = \begin{pmatrix}0\\ \mathbf{v}\end{pmatrix} \quad \text{and} \quad \mathbf{L}\begin{pmatrix}\bar{\mathbf{y}}\\ \bar{\mathbf{p}}\end{pmatrix} = \begin{pmatrix}0\\ \mathbf{w}\end{pmatrix} \tag{6.6.15}$$

respectively.

Recall now the following lemma on the convergence of gradient methods, see e.g. [Ci2] for a proof.

Lemma 6.21 *Let $\bar{\mathbf{J}}$ be a functional on ℓ_2. Suppose that there are two constants c_*, C_* such that*

(i) $\bar{\mathbf{J}}$ has a local minimum at $\mathbf{u}^*$ and is twice differentiable in an open ball $B_r(\mathbf{u}^*)$ around $\mathbf{u}^*$;

(ii) for all $\mathbf{u}\in B_r(\mathbf{u}^*)$ the functional $\bar{\mathbf{J}}$ satisfies

$$D^2\bar{\mathbf{J}}(\mathbf{u};\mathbf{v},\mathbf{w}) \;\le\; C_*\|\mathbf{v}\|_{\ell_2}\|\mathbf{w}\|_{\ell_2} \tag{6.6.16}$$

for all $\mathbf{v},\mathbf{w}\in\ell_2$;

(iii) for all $\mathbf{u}\in B_r(\mathbf{u}^*)$ and for every $\mathbf{v}\in\ell_2$ one has

$$D^2\bar{\mathbf{J}}(\mathbf{u};\mathbf{v},\mathbf{v}) \;\ge\; c_*\|\mathbf{v}\|_{\ell_2}^2. \tag{6.6.17}$$

Let $\mathbf{u}^{(0)}\in B_r(\mathbf{u}^*)$ be some initial guess and let ρ_i be such that for some fixed chosen parameters ρ_*, ρ^* one has

$$0<\rho_* \;\le\; \rho_i \;\le\; \rho^* < 2\frac{c_*}{C_*{}^2}. \tag{6.6.18}$$

Then the sequence $\{\mathbf{u}^{(i)}\}_i$ generated by the gradient algorithm

$$\mathbf{u}^{(i+1)} \;:=\; \mathbf{u}^{(i)} - \rho_i\,\delta\bar{\mathbf{J}}(\mathbf{u}^{(i)}) \tag{6.6.19}$$

converges to $\mathbf{u}^*$. Furthermore, if $B_r(\mathbf{u}^*)=\ell_2$ and $\bar{\mathbf{J}}$ is strictly convex, then (6.6.19) converges for any initial vector $\mathbf{u}^{(0)}$.

For the functional $\mathbf{J}$ from (6.3.7), we can prove now the following.

Remark 6.22 *The functional* $\mathbf{J}$ *defined in (6.3.7) satisfies the conditions in Lemma 6.21 with constants* c_* *and* C_* *in (6.6.18) defined by (6.6.23) and (6.6.25) below.*

Proof: The proof is given here explicitly to bring out the precise role of the constants. Using the identities (6.6.14), one can estimate $D^2\mathbf{J}$ by

$$\begin{aligned} D^2\mathbf{J}(\mathbf{u};\mathbf{v},\mathbf{w}) &= \omega\langle \mathbf{Z}\tilde{\mathbf{y}}, \mathbf{Z}\bar{\mathbf{y}}\rangle + \langle \mathbf{v},\mathbf{w}\rangle \\ &\leq \omega\|\mathbf{Z}\tilde{\mathbf{y}}\|_{\ell_2}\|\mathbf{Z}\bar{\mathbf{y}}\|_{\ell_2} + \|\mathbf{v}\|_{\ell_2}\|\mathbf{w}\|_{\ell_2}. \end{aligned} \tag{6.6.20}$$

In the case of Remark 6.4(i), $\|\mathbf{Z}\tilde{\mathbf{y}}\|_{\ell_2} = \|\tilde{\mathbf{y}}\|_{\ell_2}$. In the other case (ii), one has by (6.2.2)

$$\|\mathbf{Z}\tilde{\mathbf{y}}\|_{\ell_2} \leq \mathbf{c}_{T1}\|\tilde{\mathbf{y}}\|_{\ell_2} \tag{6.6.21}$$

with some constant $\mathbf{c}_{T1}$ replacing the constant $c_{T_1,y}$ from (6.2.2). Furthermore, the isomorphism relation (4.1.22) yields

$$\|\tilde{\mathbf{y}}\|_{\ell_2} \leq \mathbf{c}_{\mathbf{L}}^{-1}\|\mathbf{v}\|_{\ell_2} \tag{6.6.22}$$

for the solution $\tilde{\mathbf{y}}$ of (6.6.15), and correspondingly for $\bar{\mathbf{y}}$. Thus, one obtains from (6.6.20) and (6.6.21) followed by (6.6.22)

$$\begin{aligned} D^2\mathbf{J}(\mathbf{u};\mathbf{v},\mathbf{w}) &\leq \left(\omega(\mathbf{c}_{T1}\mathbf{c}_{\mathbf{L}}^{-1})^2 + 1\right)\|\mathbf{v}\|_{\ell_2}\|\mathbf{w}\|_{\ell_2} \\ &=: C_*\|\mathbf{v}\|_{\ell_2}\|\mathbf{w}\|_{\ell_2}. \end{aligned} \tag{6.6.23}$$

As for the lower estimate, one trivially has

$$D^2\mathbf{J}(\mathbf{u};\mathbf{v},\mathbf{v}) \geq \omega\|\mathbf{Z}\tilde{\mathbf{y}}\|_{\ell_2}^2 + \|\mathbf{v}\|_{\ell_2}^2 \;\geq\; \|\mathbf{v}\|_{\ell_2}^2, \tag{6.6.24}$$

i.e., (6.6.17) is satisfied with

$$c_* := 1. \tag{6.6.25}$$

∎

Remark 6.22 combined with Lemma 6.21 immediately yields the following fact.

Proposition 6.23 Algorithm Basic *converges for* ρ_i *satisfying*

$$0 < \rho_* \leq \rho_i \leq \rho^* < 2\frac{c_*}{(C_*)^2}, \tag{6.6.26}$$

where c_* *and* C_* *are defined in (6.6.23) and (6.6.25).*

The next result is the first step into making Algorithm Basic computationally more efficient.

Proposition 6.24 *In terms of the iterates produced by* Algorithm Basic, $\delta\mathbf{J}(\mathbf{u}^{(i)})$ *has the representation (6.6.8),*

$$\delta\mathbf{J}(\mathbf{u}^{(i)}) \;=\; \mathbf{u}^{(i)} - \boldsymbol{\mu}^{(i+1)},$$

that is, the evaluation of $\delta\mathbf{J}(\mathbf{u}^{(i)})$ *is equivalent to solving first (6.6.6) and then (6.6.7).*

Proof: Recall from (6.3.10) that

$$\delta \mathbf{J}(\mathbf{u}) \;=\; \omega \widetilde{\mathbf{B}} \mathbf{A}^{-T} \mathbf{Z}^T (\mathbf{Z}\,\mathbf{y}(\mathbf{u}) - \mathbf{y}_\Box) + \mathbf{u}.$$

Now let $\mathbf{y}^{(i+1)}(\mathbf{u}^{(i)})$ be the solution of (6.6.6). Then we obtain from (6.3.10) the representation

$$\delta \mathbf{J}(\mathbf{u}^{(i)}) = \omega \widetilde{\mathbf{B}} \mathbf{A}^{-T} \mathbf{Z}^T (\mathbf{Z}\,\mathbf{y}^{(i+1)}(\mathbf{u}^{(i)}) - \mathbf{y}_\Box) + \mathbf{u}^{(i)}. \tag{6.6.27}$$

Writing the first equation in (6.6.7) explicitly in terms of $\mathbf{z}^{(i+1)}$ gives

$$\mathbf{z}^{(i+1)} = -\omega \mathbf{A}^{-T} \mathbf{Z}^T (\mathbf{Z}\,\mathbf{y}^{(i+1)}(\mathbf{u}^{(i)}) - \mathbf{y}_\Box) - \mathbf{A}^{-T} \mathbf{B}^T \boldsymbol{\mu}^{(i+1)} \tag{6.6.28}$$

which inserted into the second equation in (6.6.7), $\mathbf{B}\mathbf{z}^{(i+1)} = \mathbf{0}$, yields

$$\omega \mathbf{B} \mathbf{A}^{-T} \mathbf{Z}^T (\mathbf{Z}\,\mathbf{y}^{(i+1)}(\mathbf{u}^{(i)}) - \mathbf{y}_\Box) + \mathbf{B} \mathbf{A}^{-T} \mathbf{B}^T \boldsymbol{\mu}^{(i+1)} = \mathbf{0}. \tag{6.6.29}$$

The latter relation can be equivalently written as

$$\boldsymbol{\mu}^{(i+1)} = -\omega (\mathbf{B} \mathbf{A}^{-T} \mathbf{B}^T)^{-1} \mathbf{B} \mathbf{A}^{-T} \mathbf{Z}^T (\mathbf{Z}\,\mathbf{y}^{(i+1)}(\mathbf{u}^{(i)}) - \mathbf{y}_\Box) \tag{6.6.30}$$

which is just

$$\boldsymbol{\mu}^{(i+1)} = -\omega \widetilde{\mathbf{B}} \mathbf{A}^{-T} \mathbf{Z}^T (\mathbf{Z}\,\mathbf{y}^{(i+1)}(\mathbf{u}^{(i)}) - \mathbf{y}_\Box) \tag{6.6.31}$$

by recalling the definition (6.3.6) of $\widetilde{\mathbf{B}}$. In view of (6.6.27), we therefore have

$$\begin{aligned} \delta \mathbf{J}(\mathbf{u}^{(i)}) &= \omega \widetilde{\mathbf{B}} \mathbf{A}^{-T} \mathbf{Z}^T (\mathbf{Z}\,\mathbf{y}^{(i+1)}(\mathbf{u}^{(i)}) - \mathbf{y}_\Box) + \mathbf{u}^{(i)} \\ &= -\boldsymbol{\mu}^{(i+1)} + \mathbf{u}^{(i)}. \end{aligned} \tag{6.6.32}$$

which confirms (6.6.8). Note that by recalling the form of $\mathbf{y}$ from (6.3.6), one can write $\mathbf{J}(\mathbf{u}^{(i)})$ explicitly in terms of $\mathbf{u}^{(i)}$ as

$$\begin{aligned} \delta \mathbf{J}(\mathbf{u}^{(i)}) &= \omega \widetilde{\mathbf{B}} \mathbf{A}^{-T} \mathbf{Z}^T \left(\mathbf{Z} \mathbf{A}^{-1} (\widetilde{\mathbf{B}}^T \mathbf{u}^{(i)} + \widetilde{\mathbf{f}}) - \mathbf{y}_\Box \right) + \mathbf{u}^{(i)} \\ &=: \widehat{\mathbf{A}} \mathbf{u}^{(i)} + \widehat{\mathbf{f}} \end{aligned} \tag{6.6.33}$$

where $\widehat{\mathbf{A}}$ is symmetric positive definite. ■

Recall from e.g. [Br] that the convergence speed θ_{grad} of the gradient method (6.6.5) is governed by the spectral condition number of $\widehat{\mathbf{A}}$,

$$\theta_{\text{grad}} = \frac{\kappa(\widehat{\mathbf{A}}) - 1}{\kappa(\widehat{\mathbf{A}}) + 1}. \tag{6.6.34}$$

Due to the preconditioning and scaling of the ingredients of $\mathbf{L}$, $\widehat{\mathbf{A}}$ can be shown to have uniformly bounded condition numbers such that

$$\theta \leq \theta_{\text{grad}} \;<\; 1 \tag{6.6.35}$$

holds independent of the discretization. Indexing now again the finite–dimensional vectors by Λ, this means that in each iteration of the gradient method the error will be reduced by a fixed fraction θ, i.e.,

$$\|\mathbf{u}_\Lambda^{(i+1)} - \mathbf{u}_\Lambda\|_{\ell_2} \leq \theta \|\mathbf{u}_\Lambda^{(i)} - \mathbf{u}_\Lambda\|_{\ell_2} \tag{6.6.36}$$

where $\mathbf{u}_\Lambda$ is the exact solution of minimizing $\mathbf{J}(\mathbf{u}_\Lambda)$ from (6.3.7) over (6.6.2).

Now we derive conditions under which the gradient scheme (6.6.13) converges, depending on the tolerances $\varepsilon_y(i+1)$ and $\varepsilon_\mu(i+1)$ from (6.6.11) and (6.6.12) up to which the two saddle point problems are solved. To this end, we denote the exact solutions $\mathbf{y}_\Lambda^{(i+1)}$ and $\boldsymbol{\mu}_\Lambda^{(i+1)}$ of (6.6.6) and (6.6.7), respectively, by

$$\mathbf{y}_\Lambda^{(i+1)} = \mathbf{y}(\mathbf{u}_\Lambda^{(i)}), \qquad \boldsymbol{\mu}_\Lambda^{(i+1)} = \boldsymbol{\mu}(\mathbf{y}_\Lambda^{(i+1)}), \tag{6.6.37}$$

emphasizing only the dependence on these variables that are relevant for the subsequent analysis. In this notation, (6.6.5) takes on the form

$$\mathbf{u}_\Lambda^{(i+1)} = \mathbf{u}_\Lambda^{(i)} + \rho_i \left(\mathbf{u}_\Lambda^{(i)} - \boldsymbol{\mu}(\mathbf{y}(\mathbf{u}_\Lambda^{(i)}))\right). \tag{6.6.38}$$

On the other hand, the iterative solution of (6.6.6) and (6.6.7) yields approximations

$$\widetilde{\mathbf{y}}_\Lambda^{(i+1)} \approx \mathbf{y}(\mathbf{u}_\Lambda^{(i)}), \qquad \widetilde{\boldsymbol{\mu}}_\Lambda^{(i+1)} \approx \boldsymbol{\mu}(\widetilde{\mathbf{y}}_\Lambda^{(i+1)}), \tag{6.6.39}$$

respectively. The iteration (6.6.13) in ALGORITHM OUTIT is executed using $\widetilde{\boldsymbol{\mu}}_\Lambda^{(i+1)}$ instead of $\boldsymbol{\mu}_\Lambda^{(i+1)}$. Adding zeroes, we can rewrite (6.6.13) in terms of $\mathbf{J}(\mathbf{u}_\Lambda^{(i)})$ as

$$\begin{aligned}
\mathbf{u}_\Lambda^{(i+1)} &:= \mathbf{u}_\Lambda^{(i)} - \rho_i \left(\mathbf{u}_\Lambda^{(i)} - \widetilde{\boldsymbol{\mu}}_\Lambda^{(i+1)}\right) \\
&= \mathbf{u}_\Lambda^{(i)} - \rho_i \Big(\mathbf{u}_\Lambda^{(i)} - \boldsymbol{\mu}(\mathbf{y}(\mathbf{u}_\Lambda^{(i)})) \\
&\qquad + \boldsymbol{\mu}(\mathbf{y}(\mathbf{u}_\Lambda^{(i)})) - \boldsymbol{\mu}(\widetilde{\mathbf{y}}_\Lambda^{(i+1)}) + \boldsymbol{\mu}(\widetilde{\mathbf{y}}_\Lambda^{(i+1)}) - \widetilde{\boldsymbol{\mu}}_\Lambda^{(i+1)}\Big) \\
&= \mathbf{u}_\Lambda^{(i)} - \rho_i \, \mathbf{J}(\mathbf{u}_\Lambda^{(i)}) \\
&\qquad + \rho_i \left(\boldsymbol{\mu}(\mathbf{y}(\mathbf{u}_\Lambda^{(i)})) - \boldsymbol{\mu}(\widetilde{\mathbf{y}}_\Lambda^{(i+1)}) + \boldsymbol{\mu}(\widetilde{\mathbf{y}}_\Lambda^{(i+1)}) - \widetilde{\boldsymbol{\mu}}_\Lambda^{(i+1)}\right) \\
&= \widehat{\mathbf{u}}_\Lambda^{(i+1)} + \rho_i \left(\boldsymbol{\mu}(\mathbf{y}(\mathbf{u}_\Lambda^{(i)})) - \boldsymbol{\mu}(\widetilde{\mathbf{y}}_\Lambda^{(i+1)}) + \boldsymbol{\mu}(\widetilde{\mathbf{y}}_\Lambda^{(i+1)}) - \widetilde{\boldsymbol{\mu}}_\Lambda^{(i+1)}\right),
\end{aligned} \tag{6.6.40}$$

where $\widehat{\mathbf{u}}_\Lambda^{(i+1)}$ is the solution of the exact gradient step (6.6.5). Recalling that the gradient method (6.6.5) satisfies the error reduction estimate (6.6.36) with fixed $\theta < 1$, we obtain by inserting (6.6.40)

$$\begin{aligned}
\|\mathbf{u}_\Lambda^{(i+1)} - \mathbf{u}_\Lambda\|_{\ell_2} &\leq \|\widehat{\mathbf{u}}_\Lambda^{(i+1)} - \mathbf{u}_\Lambda\|_{\ell_2} + \|\mathbf{u}_\Lambda^{(i+1)} - \widehat{\mathbf{u}}_\Lambda^{(i+1)}\|_{\ell_2} \qquad (6.6.41) \\
&\leq \theta \|\mathbf{u}_\Lambda^{(i)} - \mathbf{u}_\Lambda\|_{\ell_2} + \rho_i \|\boldsymbol{\mu}(\mathbf{y}(\mathbf{u}_\Lambda^{(i)})) - \boldsymbol{\mu}(\widetilde{\mathbf{y}}_\Lambda^{(i+1)}) + \boldsymbol{\mu}(\widetilde{\mathbf{y}}_\Lambda^{(i+1)}) - \widetilde{\boldsymbol{\mu}}_\Lambda^{(i+1)}\|_{\ell_2} \\
&\leq \theta \|\mathbf{u}_\Lambda^{(i)} - \mathbf{u}_\Lambda\|_{\ell_2} + \rho_i \Big(\mathbf{c}_{\mathbf{L}}^{-1} \|\mathbf{y}(\mathbf{u}_\Lambda^{(i)}) - \widetilde{\mathbf{y}}_\Lambda^{(i+1)}\|_{\ell_2} \\
&\qquad + \|\boldsymbol{\mu}(\widetilde{\mathbf{y}}_\Lambda^{(i+1)}) - \widetilde{\boldsymbol{\mu}}_\Lambda^{(i+1)}\|_{\ell_2}\Big)
\end{aligned}$$

where we have used also that $\mathbf{L}_\Lambda$ is an isomorphism with constants given in (4.1.22). Recalling the upper estimate in (4.1.22), this yields together with (6.6.11) and (6.6.12)

$$\begin{aligned} \|\mathbf{u}_\Lambda^{(i+1)} - \mathbf{u}_\Lambda\|_{\ell_2} &\leq \theta\,\|\mathbf{u}_\Lambda^{(i)} - \mathbf{u}_\Lambda\|_{\ell_2} + \rho_i\left(\mathbf{c}_\mathbf{L}^{-1}\mathbf{C}_\mathbf{L}\varepsilon_y(i+1) + \mathbf{C}_\mathbf{L}\varepsilon_\mu(i+1)\right) \\ &= \theta\,\|\mathbf{u}_\Lambda^{(i)} - \mathbf{u}_\Lambda\|_{\ell_2} + e_i \qquad (6.6.42) \end{aligned}$$

where

$$e_i := \rho_i\left(\mathbf{c}_\mathbf{L}^{-1}\mathbf{C}_\mathbf{L}\varepsilon_y(i+1) + \mathbf{C}_\mathbf{L}\varepsilon_\mu(i+1)\right). \qquad (6.6.43)$$

Repeating this argument leads to the recursion

$$\|\mathbf{u}_\Lambda^{(i+1)} - \mathbf{u}_\Lambda\|_{\ell_2} \leq \theta^{i+1}\,\|\mathbf{u}_\Lambda^{(0)} - \mathbf{u}_\Lambda\|_{\ell_2} + \sum_{l=0}^{i}\theta^l\,e_{i-l}. \qquad (6.6.44)$$

Thus, when e_i satisfies e.g.

$$e_i \leq \frac{\theta^i}{(1+i)^2}\,\tau(\mathbf{u}_\Lambda), \qquad (6.6.45)$$

where

$$\|\mathbf{u}_\Lambda^{(0)} - \mathbf{u}_\Lambda\|_{\ell_2} \leq \tau(\mathbf{u}_\Lambda), \qquad (6.6.46)$$

(where $\tau(\mathbf{u}_\Lambda)$ is an estimate for the discretization error with respect to Λ), one can conclude from (6.6.44)

$$\begin{aligned} \|\mathbf{u}_\Lambda^{(i+1)} - \mathbf{u}_\Lambda\|_{\ell_2} &\leq \theta^{i+1}\,\tau(\mathbf{u}_\Lambda) + \theta^i\tau(\mathbf{u}_\Lambda)\sum_{l=0}^{i}\frac{1}{(1+l-i)^2} \\ &=: \theta^{i+1}\,\tau(\mathbf{u}_\Lambda) + \theta^i\,\tau(\mathbf{u}_\Lambda)\,\bar{c} \\ &\leq \theta^i\,\tau(\mathbf{u}_\Lambda)\,(\theta + \bar{c}). \qquad (6.6.47) \end{aligned}$$

Summarizing, we have proved convergence of the fully iterative scheme ALGORITHM OUTIT under the condition that e_i satisfies (6.6.45). This is the case if for instance at the $(i+1)$th stage the tolerances $\varepsilon_y(i+1)$, $\varepsilon_\mu(i+1)$ are chosen as

$$\begin{aligned} \varepsilon_y(i+1) &:= \frac{1}{2}\,\frac{\mathbf{c}_\mathbf{L}}{\mathbf{C}_\mathbf{L}\,\rho_i}\,\frac{\theta^i}{(1+i)^2}\,\tau(\mathbf{u}_\Lambda), \\ \varepsilon_\mu(i+1) &:= \frac{1}{2}\,\frac{1}{\mathbf{C}_\mathbf{L}\,\rho_i}\,\frac{\theta^i}{(1+i)^2}\,\tau(\mathbf{u}_\Lambda). \end{aligned} \qquad (6.6.48)$$

Theorem 6.25 *If the tolerances $\varepsilon_y(i+1)$ and $\varepsilon_\mu(i+1)$ in (6.6.11) and (6.6.12) are selected at each stage according to (6.6.48) then* ALGORITHM OUTIT *converges for ρ_i satisfying (6.6.26),*

$$0 < \rho_* \leq \rho_i \leq \rho^* < 2\frac{c_*}{{C_*}^2},$$

where c_* and C_* are defined in (6.6.23) and (6.6.25).

In the next subsection, a detailed complexity analysis yields that this *basic iterative scheme* leads in combination with a nested iteration strategy to an *asymptotically optimal* method.

6.6.4 Computational Work for Algorithm OutIt

Up to this point, we have not specified the particular iterative method Algorithm InnIt by which (6.6.2) and (6.6.3) are solved. A simple iterative method for saddle point problems for symmetric $\mathbf{A}_\Lambda$ is the *Uzawa algorithm.* For a system of the form

$$\mathbf{L}_\Lambda \begin{pmatrix} \mathbf{y}_\Lambda \\ \mathbf{p}_\Lambda \end{pmatrix} \equiv \begin{pmatrix} \mathbf{A}_\Lambda & \mathbf{B}_\Lambda^T \\ \mathbf{B}_\Lambda & \mathbf{0} \end{pmatrix} \begin{pmatrix} \mathbf{y}_\Lambda \\ \mathbf{p}_\Lambda \end{pmatrix} = \begin{pmatrix} \mathbf{f}_\Lambda \\ \mathbf{g}_\Lambda \end{pmatrix}, \tag{6.6.49}$$

the Uzawa algorithm reads in its simplest form for $i = 0, 1, \ldots$ when $\mathbf{y}_\Lambda^{(i)}, \mathbf{p}_\Lambda^{(i)}$ are chosen,

$$\begin{aligned} \mathbf{y}_\Lambda^{(i+1)} &= \mathbf{A}_\Lambda^{-1}(\mathbf{f}_\Lambda - \mathbf{B}_\Lambda^T \mathbf{p}_\Lambda^{(i)}) \\ &= \mathbf{y}_\Lambda^{(i)} + \mathbf{A}_\Lambda^{-1}(\mathbf{f}_\Lambda - \mathbf{A}_\Lambda \mathbf{y}_\Lambda^{(i)} - \mathbf{B}_\Lambda^T \mathbf{p}_\Lambda^{(i)}) \\ \mathbf{p}_\Lambda^{(i+1)} &= \mathbf{p}_\Lambda^{(i)} + \gamma (\mathbf{B}_\Lambda \mathbf{y}_\Lambda^{(i+1)} - \mathbf{g}_\Lambda). \end{aligned} \tag{6.6.50}$$

Here γ is some sufficiently small fixed step size parameter. The first system in (6.6.50) is not solved exactly. Its iterative solution by e.g. the conjugate gradient method corresponds to applying some approximation $(\mathbf{A}_\Lambda)_0^{-1}$ of $\mathbf{A}_\Lambda^{-1}$ which can be viewed as a preconditioner for $\mathbf{A}_\Lambda$. One usually also includes a preconditioner $(\mathbf{S}_\Lambda)_0$ for the second equation,

$$\begin{aligned} \mathbf{y}_\Lambda^{(i+1)} &= \mathbf{y}_\Lambda^{(i)} + (\mathbf{A}_\Lambda)_0^{-1}(\mathbf{f}_\Lambda - \mathbf{A}_\Lambda \mathbf{y}_\Lambda^{(i)} - \mathbf{B}_\Lambda^T \mathbf{p}_\Lambda^{(i)}) \\ \mathbf{p}_\Lambda^{(i+1)} &= \mathbf{p}_\Lambda^{(i)} + \gamma (\mathbf{S}_\Lambda)_0^{-1} (\mathbf{B}_\Lambda \mathbf{y}_\Lambda^{(i+1)} - \mathbf{g}_\Lambda). \end{aligned} \tag{6.6.51}$$

The role of $(\mathbf{S}_\Lambda)_0$ is explained below. Algorithm (6.6.51) is often called *incomplete Uzawa algorithm* since the iterative method for the first equation corresponds to multiplying by an approximation $(\mathbf{A}_\Lambda)_0^{-1}$ of $\mathbf{A}_\Lambda^{-1}$, see [BPV1].

For discussing the convergence properties of (6.6.50), one considers the *reduced equation*

$$\mathbf{B}_\Lambda \mathbf{A}_\Lambda^{-1} \mathbf{B}_\Lambda^T \mathbf{p}_\Lambda = \mathbf{B}_\Lambda \mathbf{A}_\Lambda^{-1} \mathbf{f}_\Lambda - \mathbf{g}_\Lambda \tag{6.6.52}$$

involving the *Schur complement* of (6.6.49). For symmetric and positive definite $\mathbf{A}_\Lambda$, the Uzawa method (6.6.50) is known to converge if $\mathbf{B}_\Lambda \mathbf{A}_\Lambda^{-1} \mathbf{B}_\Lambda^T$ is symmetric positive definite and if e.g. the step size parameter γ satisfies

$$\gamma \; < \; 2 \, \|\mathbf{B}_\Lambda \mathbf{A}_\Lambda^{-1} \mathbf{B}_\Lambda^T\|^{-1}, \tag{6.6.53}$$

see e.g. [DHU] for a detailed derivation. In fact, an iteration for (6.6.52) reads

$$\mathbf{p}_\Lambda^{(i+1)} = (\mathbf{I} - \gamma \mathbf{B}_\Lambda \mathbf{A}_\Lambda^{-1} \mathbf{B}_\Lambda^T)\, \mathbf{p}_\Lambda^{(i)} + \mathbf{B}_\Lambda \mathbf{A}_\Lambda^{-1} \mathbf{f}_\Lambda - \mathbf{g}_\Lambda \tag{6.6.54}$$

which converges if

$$\|\mathbf{I} - \gamma \mathbf{B}_\Lambda \mathbf{A}_\Lambda^{-1} \mathbf{B}_\Lambda^T\| < 1 \tag{6.6.55}$$

which follows from (6.6.53). In the present situation, the preconditioner $\mathbf{S}_0$ is actually only needed for a possibly diagonal scaling since the Schur complement already has a uniformly bounded condition number, see the proof of Remark 6.26 below.

For the systems (6.6.6) and (6.6.7) which satisfy Corollary 6.16, we can say the following.

Remark 6.26 *The convergence rate of solving (6.6.6) or (6.6.7) by the incomplete Uzawa algorithm (6.6.51) is for suitable choices of* $(\mathbf{A}_\Lambda)_0, (\mathbf{S}_\Lambda)_0$ *independent of the discretization.*

Proof: The convergence rate θ_{Uz} of the Uzawa algorithm (6.6.50) (using conjugate gradient iterations for equation (6.6.54) and computing $\mathbf{A}_\Lambda^{-1}$ exactly) is governed by the spectral condition number of the Schur complement,

$$\theta_{\mathrm{Uz}} = \frac{\sqrt{\kappa(\mathbf{B}_\Lambda \mathbf{A}_\Lambda^{-1} \mathbf{B}_\Lambda^T)} - 1}{\sqrt{\kappa(\mathbf{B}_\Lambda \mathbf{A}_\Lambda^{-1} \mathbf{B}_\Lambda^T)} + 1}, \tag{6.6.56}$$

see e.g. [Br]. A proper scaling as in Section 6.3 yields that $\mathbf{B}_\Lambda \mathbf{A}_\Lambda^{-1} \mathbf{B}_\Lambda^T$ is an ℓ_2–automorphism such that θ_{Uz} is independent of the refinement level. From the analysis in [BPV1] one has that the convergence rate for the incomplete Uzawa algorithm (6.6.51) satisfies

$$\theta_{\mathrm{iUz}} \;=\; \frac{\delta_1(1-\delta_2) + \sqrt{\delta_1^2(1-\delta_2)^2 + 4\,\delta_2}}{2}, \tag{6.6.57}$$

where $0 < \delta_1, \delta_2 < 1$ are the convergence rates for each of the iterations in (6.6.51). Since $\mathbf{A}_\Lambda$ and $\mathbf{B}_\Lambda \mathbf{A}_\Lambda^{-1} \mathbf{B}_\Lambda^T$ have uniformly bounded condition numbers, the action of $(\mathbf{A}_\Lambda)_0^{-1}, (\mathbf{S}_\Lambda)_0^{-1}$ only means a scaling by a constant independent of the discretization such that indeed the convergence rates satisfy $\delta_1, \delta_2 < 1$. Thus, the convergence rate θ_{iUz} is also independent of the discretization level. Furthermore, since θ_{iUz} can be estimated as

$$\theta_{\mathrm{iUz}} \;\le\; 1 - \tfrac{1}{2}(1-\delta_1)(1-\delta_2)$$

it follows that

$$\theta_{\mathrm{iUz}} \;< 1.$$

∎

Consequently, choosing the incomplete Uzawa method (6.6.51) as inner iteration ALGORITHM INNIT in ALGORITHM OUTIT, in both STEP 2 and STEP 3 for any size of the systems (6.6.11), (6.6.12) only a *fixed* number of iterations is needed to reduce the error by a fixed fraction. Recall also that each iteration can be applied in an amount of work proportional to the size of the system since all operators in $\mathbf{L}_\Lambda$ can be realized by successively applying sparse matrices, see Section 5.6.

In the following, we exploit the combination of the basic iterative method ALGORITHM OUTIT and the analysis used to prove Theorem 6.25 with a nested iteration strategy to derive an asymptotically optimal algorithm. The reasoning is in principle, similar to that in Sections 5.6 and 6.6.3 although the derivation is somewhat more involved. In particular, the accuracy of the iterative solution of the two systems has to be made dependent also on the discretization error for $\mathbf{u}$.

The derivation will be detailed for uniform refinements so that we exchange the index Λ again by the parameter j denoting some discretization level. Denote by $\tau(\mathbf{u}_j)$ an estimate for the *discretization error* for the control on level j, i.e.,

$$\|\mathbf{u}_j - \mathbf{u}\|_{\ell_2} \;\le\; \tau(\mathbf{u}_j), \tag{6.6.58}$$

where $\mathbf{u}$ is the exact solution of the infinite–dimensional system (6.3.5) and $\mathbf{u}_j$ is the exact solution of (6.6.3) on level j. Suppose that we have determined on level j an approximation $\mathbf{u}_j^{(i)}$ such that

$$\|\mathbf{u}_j^{(i)} - \mathbf{u}_j\|_{\ell_2} \;\leq\; \tfrac{1}{2}\tau(\mathbf{u}_j). \tag{6.6.59}$$

Now choose $\mathbf{u}_j^{(i)}$ as initial guess for the iteration (6.6.13) on level $j+1$. Note that since $\mathbf{u}_j^{(i)}$ consists of wavelet coefficients we can formally set

$$\mathbf{u}_{j+1}^{(0)} := \mathbf{u}_j^{(i)}, \tag{6.6.60}$$

meaning that the array $\mathbf{u}_j^{(i)}$ is extended to the vector of wavelet coefficients on level $j+1$ by simply appending zeros. We will determine next a number $i+1 = i_{j+1}$ of gradient iterations which is needed to reduce the initial error

$$\|\mathbf{u}_{j+1}^{(0)} - \mathbf{u}_{j+1}\|_{\ell_2}$$

on level $j+1$ so that

$$\|\mathbf{u}_{j+1}^{(i_{j+1})} - \mathbf{u}_{j+1}\|_{\ell_2} \;\leq\; \frac{\tau(\mathbf{u}_{j+1})}{2}, \tag{6.6.61}$$

which would advance (6.6.59). Recall from (6.6.44) and (6.6.47) that one has

$$\|\mathbf{u}_{j+1}^{(i+1)} - \mathbf{u}_{j+1}\|_{\ell_2} \;\leq\; \theta^{i+1}\,\|\mathbf{u}_{j+1}^{(0)} - \mathbf{u}_{j+1}\|_{\ell_2} + \theta^i\bar{c}\,\tau(\mathbf{u}_j), \tag{6.6.62}$$

provided that the bounds e_i satisfy now the following counterpart to (6.6.45),

$$e_i \;\leq\; \frac{\theta^i}{(1+i)^2}\,\tau(\mathbf{u}_j). \tag{6.6.63}$$

Moreover, from (6.6.59), (6.6.60) and (6.6.58) we infer

$$\begin{aligned}\|\mathbf{u}_{j+1}^{(0)} - \mathbf{u}_{j+1}\|_{\ell_2} \;&\leq\; \|\mathbf{u}_j^{(i)} - \mathbf{u}_j\|_{\ell_2} + \|\mathbf{u}_j - \mathbf{u}\|_{\ell_2} + \|\mathbf{u}_{j+1} - \mathbf{u}\|_{\ell_2} \\ &\leq\; \tfrac{1}{2}\tau(\mathbf{u}_j) + \tau(\mathbf{u}_j) + \tau(\mathbf{u}_{j+1}) \\ &=\; \tfrac{3}{2}\tau(\mathbf{u}_j) + \tau(\mathbf{u}_{j+1}).\end{aligned} \tag{6.6.64}$$

At this point, we need to recall some facts about the underlying framework. The discrete solutions $\mathbf{u}_j$ belong to spaces S_j which in turn form an ascending sequence (3.2.2) with growing j. If this sequence corresponds to uniform refinements, standard error estimates of the form (4.1.30) for u with respect to the Q'–norm combined with Jackson estimates (3.2.70) yield that the minimal rate of convergence is determined by

$$\|\mathbf{u}_j - \mathbf{u}\|_{\ell_2} \;\lesssim\; 2^{-sj}\|\mathbf{u}\|_{\ell_2}$$

where for the present applications one has at least $0 < s < 1/2$. Thus, either

$$\frac{\tau(\mathbf{u}_{j+1})}{\tau(\mathbf{u}_j)} \;=:\; \delta \;<\; 1 \tag{6.6.65}$$

holds with δ proportional to 2^{-s}, or already $\mathbf{u} \in S_j$. In the case of non-uniform refinements, a strategy like the one proposed in [DHU] assures that (6.6.65) holds. Now we can further estimate (6.6.62) by employing (6.6.65) (6.6.64) and (6.6.65)

$$\|\mathbf{u}_{j+1}^{(i+1)} - \mathbf{u}_{j+1}\|_{\ell_2} \leq \theta^i \left(\tfrac{3\theta}{2} + \theta\delta + \bar{c}\right) \tau(\mathbf{u}_j). \tag{6.6.66}$$

Thus, choosing $i = i_{j+1} - 1$ so that

$$\frac{3\theta}{2} + \theta\delta + \bar{c} \leq \delta/2,$$

we have verified (6.6.61). In particular, the numbers i_{j+1} can be kept uniformly bounded independent of the level j. Note that in terms of the tolerances $\varepsilon_y(i+1)$ and $\varepsilon_\mu(i+1)$ that control the accuracy of the inner iterations in (6.6.11) and (6.6.12) one needs, in view of (6.6.63),

$$\begin{aligned} \varepsilon_y(i+1,j) &:= \frac{1}{2} \frac{\mathbf{c_L}}{\mathbf{C_L}\,\rho_i} \frac{\theta^i}{(i+1)^2} \tau(\mathbf{u}_j), \\ \varepsilon_\mu(i+1,j) &:= \frac{1}{2} \frac{1}{\mathbf{C_L}\,\rho_i} \frac{\theta^i}{(i+1)^2} \tau(\mathbf{u}_j). \end{aligned} \tag{6.6.67}$$

Since as mentioned above $i = i_{j+1} - 1$ remains bounded, the quotients

$$\frac{\varepsilon_y(i+1,j)}{\varepsilon_y(i,j)} \qquad \text{and} \qquad \frac{\varepsilon_\mu(i+1,j)}{\varepsilon_\mu(i,j)}$$

remain proportional to δ. Thus, choosing $\widetilde{\mathbf{y}}_j^{(i)}$ and $\widetilde{\boldsymbol{\mu}}_j^{(i)}$ as initial guesses for the computation of $\widetilde{\mathbf{y}}_{j+1}^{(i+1)}$ and $\widetilde{\boldsymbol{\mu}}_{j+1}^{(i+1)}$ in (6.6.11) and (6.6.12), only a fixed level independent error reduction is used which under the above assumptions requires only a fixed number of inner iterations independent of j.

Now we continue to argue like in Section 5.6. The systems in (6.6.11) and (6.6.12) can be solved by e.g. the incomplete Uzawa method (6.6.51) with a convergent rate independent of the discretizations, see Remark 6.26. Thus, taking as initial guess the solution from the previous level, only a uniformly bounded number of Uzawa steps is required to reduce the error by a fixed fraction which is all that is needed in order to achieve discretization error accuracy on each level. On the lowest level the system is solved exactly. As shown also in Section 5.6, the operators in (6.6.11) and (6.6.12) can be applied at the expense of computational work that remains proportional to the number of unknowns on that level. Thus, again with a geometric series argument it follows that the overall work stays proportional to the computational work required by a matrix/vector multiplication on the highest level J, that is, the total work is proportional to $\mathcal{O}(N_J)$ where N_J is the number of unknowns on the highest level. In summary, we have finally proved the following result.

Theorem 6.27 *If in each iteration of* ALGORITHM OUTIT *the systems (6.6.11) and (6.6.12) are solved up to the tolerances (6.6.67) and these solutions are taken as initial guesses for the next higher level, then* ALGORITHM OUTIT *is an asymptotically optimal method in the sense that it provides the solution up to the discretization error on level J in an overall amount of $\mathcal{O}(N_J)$ operations where N_J is the number of unknowns in (6.6.11), (6.6.12) and (6.6.13).*

It should be remarked that in the above strategy the main goal is to yield an asymptotically optimal result. To obtain a quantitatively efficient scheme for each J, more care has to be taken.

6.6.5 A Numerical Example

We close this chapter with the following numerical example. The elliptic boundary value problem that plays the role of the constraints for the control problem (6.2.6) is the problem (4.3.1) already employed in Sections 4.3 and 5.7,

$$\begin{aligned} -\Delta y + y &= 1 \quad \text{in } \Omega, \\ y &= 0 \quad \text{on } \partial\Omega. \end{aligned}$$

Recall that Ω is the open disc with radius $R = 0.5$ around the mid point $(0.5, 0.5)$,

$$\Omega = \{\mathbf{x} \in I\!R^2 : (x_1 - 1/2)^2 + (x_2 - 1/2)^2 < R\},$$

which is embedded into the fictitious domain $\Box = (0,1)^2$. In the minimization functional we have chosen the first norm in (6.3.1) to be equivalent to $H(Y) = Y = H^1(\Box)$ which corresponds to the situation considered in the functional (6.1.1). Thus, the discretized properly shifted primal and dual systems read according to (6.6.2) and (6.6.3)

$$\mathbf{L}_\Lambda \begin{pmatrix} \mathbf{y}_\Lambda \\ \mathbf{p}_\Lambda \end{pmatrix} \equiv \begin{pmatrix} \mathbf{A}_\Lambda & \mathbf{B}_\Lambda^T \\ \mathbf{B}_\Lambda & \mathbf{0} \end{pmatrix} \begin{pmatrix} \mathbf{y}_\Lambda \\ \mathbf{p}_\Lambda \end{pmatrix} = \begin{pmatrix} \mathbf{f}_\Lambda \\ \mathbf{u}_\Lambda \end{pmatrix}, \qquad \mathbf{L}_\Lambda \begin{pmatrix} \mathbf{z}_\Lambda \\ \mathbf{u}_\Lambda \end{pmatrix} = -\omega \begin{pmatrix} \mathbf{y}_\Lambda - (\mathbf{y}_\Box)_\Lambda \\ 0 \end{pmatrix}. \tag{6.6.68}$$

The complete system is solved by applying the following variant of ALGORITHM OUTIT. In STEP 3 the system (6.6.12) is solved for $\mathbf{z}_\Lambda^{(i+1)}$ and $\mathbf{u}_\Lambda^{(i+1)}$ instead of $\boldsymbol{\mu}_\Lambda^{(i+1)}$, and STEP 4 is discarded since the update of $\mathbf{u}_\Lambda^{(i+1)}$ is already performed in STEP 3. As inner iteration the Uzawa algorithm is used with a preconditioned conjugate gradient method to solve the first equation in (6.6.50) iteratively. That is, we have applied the same CG–Uzawa algorithm like in the numerical tests in Section 4.3. Again our stopping criterion is based on the ℓ_2 norm of the residual which is proportional to the error of y in H^1. The first numbers in the third column of Table 6.1 reveal the total number of CG iterations necessary to force the ℓ_2 error of the residual of the primal system to be smaller than `tol`$= \min\{2^{-j}, 2^{-\ell}\}$. The inner iterations are terminated when the residual is smaller than $0.01*$`tol`. The numbers in parentheses show the number of Uzawa iterations. Note that the numbers in the third column in Table 6.1 coincide with the numbers in the third column in Table 4.5. Recall from the discussion in Chapter 4.3 that the increase of the iteration numbers when ℓ grows relative to j is caused by a violation of the sufficient conditions for the LBB condition. In STEP 3 the adjoint system is solved for $\mathbf{z}_\Lambda^{(i+1)}$ and $\mathbf{u}_\Lambda^{(i+1)}$ up to the same tolerance `tol` for the corresponding residual of that system.

It is interesting to observe that for these tolerances the variant of ALGORITHM OUTIT *always terminates* after 1 cycle, that is, system (6.6.2) is solved in STEP 2 up to the requested tolerance, followed by the solution of the system (6.6.3) in STEP 3. Returning only once more to STEP 2 is in *all* cases sufficient to meet the required overall tolerance. For this reason, the cycle

$$\text{Step 2 — Step 3 — Step 2}$$

is called *solution cycle* for the coupled saddle point problem (6.6.1). In Table 6.1 the iteration numbers for each step of the solution cycle are listed in columns 3, 4 and 5 and are termed *1st it.*, *2nd it.* and *3rd it.* The number #it means the total number of pcg-iterations with number of Uzawa steps written behind in parentheses. The total residual for system (6.6.2) is measured as before and is of the same quantitative behavior as in Table 4.5. We observe that the iteration numbers in the 3rd iteration are always smaller than the ones from the 1st iteration. This is a consequence of the fact that the solutions from the 1st iteration are taken as initial guesses to start the 3rd iteration.

Remark 6.28 *In summary, it seems to be sufficient to solve the adjoint system for determining the control only once, sandwiched between two iterative solutions of the primal system.*

There are many variants one can think of to balance the amount of iterations needed in each step of the cycle with the necessary amount of iterations.

As for surface plots of the solutions, there is no visible difference between the solution of the control problem considered in this example and the one from Section 4.3 displayed in Figures 4.1, 4.2 and 4.3 so that we dispense with these plots here.

j	ℓ	1st it.	2nd it.	3rd it.	residual
$\Box$	Γ	STEP 2	STEP 3	STEP 2	$\|\mathbf{r}_\Lambda\|_{\ell_2}$
3	3	50(5)	21(1)	11(1)	$1.22007090e-1$
3	4	63(4)	22(1)	22(2)	$3.03893551e-2$
3	5	64(4)	23(1)	12(1)	$2.55219641e-2$
3	6	100(7)	25(1)	38(3)	$1.40530157e-2$
4	3	77(4)	28(1)	14(1)	$5.64009761e-2$
4	4	64(3)	28(1)	14(1)	$4.89317933e-2$
4	5	80(4)	30(1)	43(3)	$1.88224372e-2$
4	6	313(21)	33(1)	31(2)	$1.17742201e-2$
4	7	412(28)	55(2)	46(4)	$5.25188732e-3$
5	3	90(4)	33(1)	30(2)	$1.07381850e-2$
5	4	91(4)	34(1)	31(2)	$1.73216501e-2$
5	5	90(4)	35(1)	17(1)	$2.43685913e-2$
5	6	94(4)	38(1)	49(3)	$8.63815998e-3$
5	7	470(29)	64(2)	70(4)	$5.10956365e-3$
6	3	119(5)	58(2)	20(1)	$1.43552113e-2$
6	4	119(5)	39(1)	35(2)	$1.24542308e-2$
6	5	131(6)	40(1)	36(2)	$1.13766598e-2$
6	6	102(4)	42(1)	68(4)	$1.26067723e-2$
6	7	119(5)	70(2)	21(1)	$5.97800253e-3$

Table 6.1: Iteration numbers for the coupled saddle point problem (6.6.68) solved by ALGORITHM OUTIT using in each step the (preconditioned) CG–Uzawa method; #it: total number of PCG iterations with number of Uzawa steps in parentheses until the respective residual satisfies $\|\mathbf{r}_\Lambda\|_{\ell_2} \leq$`tol`.

6.7 Outlook into Nonlinear Problems and Adaptive Strategies

There are many investigations of optimal control problems involving nonlinear equations like the Navier–Stokes equations, see e.g. [Bo, BG, Glo, Gu, GHS1, GHS2, GL2, GM, Hei1, Hei2, Hi, HK1, HK2, IK, Kf], both on theoretical subjects as well as on problems related to the numerical solution of the resulting optimal control systems. One area of future research would be to combine their results with the ones obtained here.

Moreover, it would be interesting to combine the above ideas with *adaptive* solution strategies. Meanwhile numerous investigations of adaptive wavelet concepts have been documented in the literature. A systematic approach based on a–posteriori error estimators was considered in [Bel] for a special case and in [DDHS] for a wide class of symmetric elliptic operator equations where, in particular, convergence in the energy norm could be established rigorously. While these results still leave the question of computational complexity open, a modified scheme for this class of problems has recently been developed in [CDD1]. There estimates for convergence rates and computational complexity have been obtained that are asymptotically optimal. Roughly speaking, this means that the computational work stays proportional to the number of significant wavelet coefficients that are needed to recover the solution within any desired accuracy tolerance. Besides the development of corresponding new algorithmic ingredients and data structures in [BCDU], the numerical experiments obtained there confirm the predicted behavior.

Quite recently, there have also been attempts to extend these results to indefinite problems [DHU]. There the results from [DDHS] have been carried over to saddle point problems without establishing however any complexity estimates. Extensions of the convergence and complexity results in [CDD1] to a wider scope of problems such as saddle point problems or just those for which (1.2) or (6.2.31) hold have just been obtained in [CDD2]. The special case of adaptive methods for saddle point problems based on an Uzawa algorithm is together with an implementation currently under investigation [DDU].

References

[Ad] R.A. Adams, *Sobolev Spaces*, Academic Press, 1978.

[ADN] S. Agmon, A. Douglis, L. Nirenberg, *Estimates near the boundary for solutions of elliptic partial differential equations satisfying general boundary conditions. II*, Comm. Pure Appl. Math., 17, 1964, 35–92.

[Ag] A. Agouzal, *Analyse Numérique de Méthodes de Décomposition de Domaines. Méthodes de Domaines Fictifs avec Multiplicateurs de Lagrange*, Thése de Doctorat, Université de Pau, 1993.

[AHJP] L. Andersson, N. Hall, B. Jawerth, G. Peters, *Wavelets on closed subsets of the real line*, in: Topics in the Theory and Applications of Wavelets, L.L. Schumaker and G. Webb (eds.), Academic Press, Boston, 1994, 1–61.

[An] P. Angot, *Analysis of singular perturbations on the Brinkman problem for fictitious domain models of viscous flows*, Math. Meth. Appl. Sci. 22, 1999, 1395–1412.

[ABF] P. Angot, C.-H. Bruneau, P. Fabrie, *A penalization method to take into account obstacles in incompressible viscous flows*, Numer. Math. 61, 1999, 497–520.

[AN] V. Arnăutu, P. Neittaanmäki, *Discretization estimates for an elliptic control problem*, Numer. Funct. Anal. Optim. 19, 1998, 431–464.

[As] G. P. Astrakantsev, *Methods of fictitious domains for a second order elliptic equation with natural boundary conditions*, U.S.S.R. Comput. Math. Math. Phys. 18, 1978, 114–121.

[Ba1] I. Babuška, *The finite element method with Lagrange multipliers*, Numer. Math. 20, 1973, 179–192.

[Ba2] I. Babuška, *The finite element method with penalty*, Math. Comp. 27, 1973, 221–228.

[BWY] R.E. Bank, B.D. Welfert, H. Yserentant, *A class of iterative methods for solving saddle point problems*, Numer. Math. 56, 1990, 645–666.

[BCDU] A. Barinka, T. Barsch, Ph. Charton, A. Cohen, S. Dahlke, W. Dahmen, K. Urban, *Adaptive wavelet schemes for elliptic problems — Implementation and numerical experiments*, IGPM–Preprint #173, RWTH Aachen, June 1999, to appear in SIAM J. Sci. Comp.

[Ba] T. Barsch, *Adaptive Multiskalenverfahren für elliptische partielle Differentialgleichungen — Realisierung, Umsetzung und numerische Ergebnisse* (in German), IGPM, RWTH Aachen, PhD Thesis, 2001.

[BKU] T. Barsch, A. Kunoth, K. Urban, *Towards object oriented software tools for numerical multiscale methods for p.d.e.s using wavelets*, in: Multiscale Wavelet Methods for PDEs, W. Dahmen, A. J. Kurdila, P. Oswald (eds.), Academic Press, 1997, 383–412.

[BKR] R. Becker, H. Kapp, R. Rannacher, *Adaptive finite element methods for optimal control of partial differential equations: Basic concept*, Preprint 1998–55 IWR und SFB 359, University of Heidelberg, November 1998, to appear in: SIAM J. Contr. Optim.

[Be1] S. Bertoluzza, *A posteriori error estimates for the wavelet Galerkin method*, Appl. Math. Lett. 8, 1995, 1–6.

[Be2] S. Bertoluzza, *Stabilization by multiscale decomposition*, Appl. Math. Lett. 11, 1998, 129–134.

[Be3] S. Bertoluzza, *Wavelet stabilization of the Lagrange multiplier method*, Numer. Math. 86, 2000, 1–28.

[BCT] S. Bertoluzza, C.Canuto, A.Tabacco, *Negative norm stabilization of convection-diffusion problems*, Appl. Math. Lett. 13, 2000, 121–127.

[BeK] S. Bertoluzza, A. Kunoth, *Wavelet stabilization and preconditioning for domain decomposition*, IMA J. Numer. Anal. 20, 2000, 533–559.

[Bo] P.B. Bochev, *Least-squares methods for optimal control*, Nonlin. Anal. Theory Methods Appl. 30, 1997, 1875–1885.

[BG] P.B. Bochev, M.D. Gunzburger, *Finite element methods of least-squares type*, SIAM Rev. 40, 1998, 789–837.

[BEK] F. Bornemann, B. Erdmann, R. Kornhuber, *A posteriori error estimates for elliptic problems in two and three space dimensions*, SIAM J. Numer. Anal. 33, 1996, 1188–1204.

[Br] D. Braess, *Finite Elements: Theory, Fast Solvers and Applications in Solid Mechanics*, Cambridge University Press, 1997.

[Brl] J.H. Bramble, *The Lagrange multiplier method for Dirichlet's problem*, Math. Comp. 37, 1981, 1–11.

[BLP1] J.H. Bramble, R.D. Lazarov, J.E. Pasciak, *A least-squares approach based on a discrete minus one inner product for first order systems*, Math. Comp. 66, 1997, 935–955.

[BLP2] J.H. Bramble, R.D. Lazarov, J.E. Pasciak, *Least-squares for second order elliptic problems*, Comp. Meth. Appl. Mech. Engnrg. 152, 1998, 195–210.

[BP] J.H. Bramble, J. Pasciak, *A preconditioning technique for indefinite systems resulting from mixed approximations for elliptic problems*, Math. Comp. 50, 1988, 1–17.

[BPV1] J.H. Bramble, J. Pasciak, A.T. Vassilev, *Analysis of the inexact Uzawa algorithm for saddle point problems*, SIAM J. Num. Anal. 34, 1997, 1072–1092.

[BPV2] J.H. Bramble, J.E. Pasciak, P.S. Vassilevski, *Computational scales of Sobolev norms with application to preconditioning*, Math. Comp. 69, 2000, 443–462.

[BF] F. Brezzi, M. Fortin, *Mixed and Hybrid Finite Element Methods*, Springer, 1991.

[BM] F. Brezzi, L.D. Marini, *A three-field domain decomposition method*, Contemp. Maths. 157, 1994, 27-34.

[CLMM] Z. Cai, R. Lazarov, T.A. Manteuffel, S.F. McCormick, *First-order system least squares for second-order partial differential equations; Part I*, SIAM J. Numer. Anal. 31, 1994, 1785–1799.

[CaM] C. Canuto, R. Masson, *Stabilized wavelet approximations of the Stokes problem*, Preprint n. 99-23, Dipartimento di Matematica, Politecnico di Torino, to appear in Math. Comp.

[CTU1] C. Canuto, A. Tabacco, K. Urban, *The wavelet element method, part I: Construction and analysis*, Appl. Comput. Harm. Anal. 6, 1999, 1–52.

[CTU2] C. Canuto, A. Tabacco, K. Urban, *The wavelet element method, part II: Realization and additional features in 2D and 3D*, Appl. Comput. Harm. Anal. 8, 2000, 123–165.

[CTU3] C. Canuto, A. Tabacco, K. Urban, *Numerical solution of elliptic problems by the Wavelet Element Method*, in: ENUMATH 97, H.G. Bock et al, eds., World Scientific Co., Singapore, 1998, 17–37.

[CDP] J.M. Carnicer, W. Dahmen, J.M. Peña, *Local decomposition of refinable spaces*, Appl. Comp. Harm. Anal. 3, 1996, 127–153.

[Ci1] P. Ciarlet, *The Finite Element Method for Elliptic Problems*, North–Holland, 1978.

[Ci2] P. Ciarlet, *Introduction to Numerical Linear Algebra and Optimization*, Cambridge, 1989.

[CF] Z. Ciesielski, T. Figiel, *Spline bases in classical function spaces on compact C^∞ manifolds: Part I and II*, Studia Mathematica, 1983, 1–58 and 95–136.

[C] A. Cohen, *Numerical Analysis of Wavelet Methods*, Handbook of Numerical Analysis II, vol. 8, P.G. Ciarlet, J.L. Lions (eds.), Elsevier Science Publishers, 1998.

[CDD1] A. Cohen, W. Dahmen, R. DeVore, *Adaptive wavelet methods for elliptic operator equations — Convergence rates*, Math. Comp. 70, 2001, 27–75.

[CDD2] A. Cohen, W. Dahmen, R. DeVore, *Adaptive wavelet methods II — Beyond the elliptic case*, IGPM–Preprint #199, RWTH Aachen, November 2000.

[CDF] A. Cohen, I. Daubechies, J.-C. Feauveau, *Biorthogonal bases of compactly supported wavelets*, Comm. Pure Appl. Math. 45, 1992, 485–560.

[CM1] A. Cohen, R. Masson, *Wavelet adaptive methods for second order elliptic problems, boundary conditions and domain decomposition*, Numer. Math. 86, 2000, 193–238.

[CM2] A. Cohen, R. Masson, *Adaptive wavelet methods for second order elliptic problems, preconditioning and adaptivity*, SIAM J. Sci. Comp. 21, 1999, 1006–1026.

[DDHS] S. Dahlke, W. Dahmen, R. Hochmuth, R. Schneider, *Stable multiscale bases and local error estimation for elliptic problems*, Appl. Numer. Maths. 8, 1997, 21–47.

[DDU] S. Dahlke, W. Dahmen, K. Urban, *Adaptive wavelet methods for saddle point problems — Convergence rates*, in preparation.

[DHU] S. Dahlke, R. Hochmuth, K. Urban, *Adaptive wavelet methods for saddle point problems*, Mathematical Modelling and Numerical Analysis (M2AN) 34, 2000, 1003–1022.

[DkK] S. Dahlke, A. Kunoth, *Wavelet characterizations of function spaces on skeletons*, in: Proceedings of the International Wavelets Conference "Wavelets and Multiscale Methods", Tanger, Morocco, April 13–17, 1998, INRIA report.

[D1] W. Dahmen, *Some remarks on multiscale transformations, stability and biorthogonality*, in: Wavelets, Images and Surface Fitting, P.J. Laurent, A. Le Méhauté, L.L. Schumaker (eds.), Academic Press, 1994, 157–188.

[D2] W. Dahmen, *Stability of multiscale transformations*, J. Four. Anal. Appl. 2, 1996, 341–361.

[D3] W. Dahmen, *Wavelet and multiscale methods for operator equations*, Acta Numerica 1997, 55–228.

[D4] W. Dahmen, *Wavelet methods for PDEs — Some recent developments*, IGPM–Preprint #183, RWTH Aachen, December 1999.

[DHJK] W. Dahmen, B. Han, R.–Q. Jia, A. Kunoth, *Biorthogonal multiwavelets on the interval: Cubic Hermite splines*, Constr. Approx. 16, 2000, 221–259.

[DHaS] W. Dahmen, H. Harbrecht, R. Schneider, private communication.

[DK1] W. Dahmen, A. Kunoth, *Multilevel preconditioning*, Numer. Math. 63, 1992, 315–344.

[DK2] W. Dahmen, A. Kunoth, *Appending boundary conditions by Lagrange multipliers: Analysis the LBB condition*, IGPM-Preprint #164, Oktober 1998, revidiert Dezember 1999, erscheint in Numer. Math. (Numer. Math. Online First DOI 10.1007/s002110000223).

[DKS1] W. Dahmen, A. Kunoth, R. Schneider, *Operator equations, multiscale concepts and complexity*, in: Mathematics of Numerical Analysis: Real Number Algorithms, J. Renegar, M. Shub, S. Smale (eds.), Lectures in Applied Mathematics 32, 1996, 225–261.

[DKS2] W. Dahmen, A. Kunoth, R. Schneider, *Wavelet least square methods for boundary value problems*, IGPM–Preprint #175, RWTH Aachen, September 1999, submitted for publication.

[DKU1] W. Dahmen, A. Kunoth, K. Urban, *A wavelet Galerkin method for the Stokes problem*, Computing 56, 1996, 259–302.

[DKU2] W. Dahmen, A. Kunoth, K. Urban, *Biorthogonal spline-wavelets on the interval - Stability and moment conditions*, Appl. Comput. Harm. Anal. 6, 1999, 132–196.

[DKU3] W. Dahmen, A. Kunoth, K. Urban, *Wavelets in numerical analysis and their quantitative properties*, in: Surface Fitting and Multiresolution Methods, A. Le Méhauté, C. Rabut and L.L. Schumaker (eds.), Vanderbilt University Press, Nashville, TN, 1997, 93–130.

[DM1] W. Dahmen, C.A. Micchelli, *Banded matrices with banded inverses, II: Locally finite decomposition of spline spaces*, Constr. Appr. 9, 1993, 263–281.

[DM2] W. Dahmen, C.A. Micchelli, *Using the refinement equation for evaluating integrals of wavelets*, SIAM J. Numer. Anal. 30, 1993, 507–537.

[DPS1] W. Dahmen, S. Prößdorf, R. Schneider, *Multiscale methods for pseudodifferential equations on smooth manifolds*, in: Proceedings of the International Conference on Wavelets: Theory, Algorithms, and Applications, C.K. Chui, L. Montefusco, L. Puccio (eds.), Academic Press, 1994, 385–424.

[DPS2] W. Dahmen, S. Prößdorf, R. Schneider, *Wavelet approximation methods for pseudodifferential equations II: Matrix compression and fast solution*, Adv. Comp. Maths. 1, 1993, 259–335.

[DS1] W. Dahmen, R. Schneider, *Composite wavelet bases for operator equations*, Math. Comp. 68, 1999, 1533–1567.

[DS2] W. Dahmen, R. Schneider, *Wavelets with complementary boundary conditions — Function spaces on the cube*, Results in Mathematics 34, 1998, 255–293.

[DS3] W. Dahmen, R. Schneider, *Wavelets on manifolds I: Construction and domain decomposition*, SIAM J. Math. Anal. 31, 1999, 184–230.

[DSt] W. Dahmen, R. Stevenson, *Element-by-element construction of wavelets satisfying stability and moment conditions*, SIAM J. Numer. Anal. 37, 1999, 319–325.

[DH] J. Dankova, R. Haslinger, *Fictitious domain approach used in shape optimization: Neumann boundary condition*, in: Casas, Eduardo (ed.), Control of Partial Differential Equations and Applications, Dekker, Lect. Notes Pure Appl. Math. 174, 1996, 43–49.

[Dau] I. Daubechies, *Orthonormal bases of compactly supported wavelets*, Comm. Pure Appl. Math. 41, 1988, 909–996.

[E] T. Elfving, *Block-iterative methods for consistent and unconsistent linear equations*, Numer. Math. 35, 1980, 1–12.

[FK] S.A. Finogenov, Y.A. Kuznetsov, *Two-stage ficititious component methods for solving the Dirichlet boundary value problem*, Sov. J. Numer. Anal. Modeling 3, 1988, 301–323.

[GJL] Ph.E. Gill, L.O. Jay, M.W. Leonard, L.R. Petzold, V. Sharma, *An SQP method for the optimal control of large-scale dynamical systems*, J. Comput. Appl. Math. 120, 2000, 197–213.

[GMPS] Ph.E. Gill, W. Murray, D.B. Ponecon, M. Saunders, *Preconditioners for indefinite systems arising in optimization*, SIAM J. Matrix Anal. Appl. 13, 1992, 292–311.

[GG] V. Girault, R. Glowinski, *Error analyis of a fictitious domain method applied to a Dirichlet problem*, Japan J. Industr. Appl. Math. 12, 1995, 487–514.

[GR] V. Girault, P.-A. Raviart, *Finite Element Methods for Navier-Stokes Equations*, Springer, 1986.

[Glo] R. Glowinski, *Finite element methods for the numerical simulation of incompressible viscous flow. Introduction to the control of the Navier-Stokes equations*, Vortex dynamics and vortex methods, Proc. 21st AMS-SIAM Semin., Seattle/WA 1990, Lect. Appl. Math. 28, 1991, 219–301.

[GPP] R. Glowinski, T.W. Pan, J. Periaux, *A fictitious domain method for Dirichlet problem and application*, Comp. Meth. Appl. Mechs. Eng. 111, 1994, 282–303.

[Gr] P. Grisvard, *Elliptic Problems in Nonsmooth Domains*, Pitman, 1985.

[Gu] M.D. Gunzburger (ed.), *Flow Control*, The IMA Volumes in Mathematics and its Applications 68, Springer, 1995.

[GuH] M.D. Gunzburger, S.L. Hou, *Treating inhomogeneous boundary conditions in finite element methods and the calculation of boundary stresses*, SIAM J. Numer. Anal., 29, 1992, 390–424.

[GHS1] M.D. Gunzburger, L.S. Hou, Th. P. Svobodny, *Analysis and finite element approximation of optimal control problems for the stationary Navier–Stokes equations with Dirichlet controls*, M^2AN 25, 1991, 711–748.

[GHS2] M.D. Gunzburger, L.S. Hou, Th. P. Svobodny, *Analysis and finite element approximation of optimal control problems for the stationary Navier-Stokes equations with distributed and Neumann controls*, Math. Comp. 57, 1991, 123–151.

[GL1] M. D. Gunzburger, H. C. Lee, *Analysis, approximation, and computation of a coupled solid/fluid temperature control problem*, Comp. Meth. Appl. Mech. Engrg. 118, 1994, 133–152.

[GL2] M. D. Gunzburger, H. C. Lee, *Analysis and approximation of optimal control problems for first-order elliptic systems in three dimensions*, Appl. Math. Comput. 100, 1999, 49–70.

[GM] M. D. Gunzburger, S. Manservisi, *The velocity tracking problem for Navier-Stokes flows with bounded distributed controls*, SIAM J. Contr. Optim. 37, 1999, 1913–1945.

[Ha1] W. Hackbusch, *Fast solution of elliptic control problems*, J. Optim. Theory Appl. 31, 1980, 565–581.

[Ha2] W. Hackbusch, *Elliptic Differential Equations: Theory and Numerical Treatment*, Springer, 1992.

[Ha2] W. Hackbusch, *Iterative Solution of Large Sparse Systems of Equations*, Springer, New York, 1994.

[Has] J. Haslinger, *Fictitious domain approaches in shape optimization*, in: Borggard, Jeff (ed.) et al., Computational Methods for Optimal Design and Control, Birkhaeuser, Prog. Syst. Control Theory 24, 1998, 237–248.

[HHK] J. Haslinger, K.–H. Hoffmann, M. Kočvara, *Control fictitious domain method for solving optimal shape design problems*, RAIRO, Modelisation Math. Anal. Numer. 27, 1993, 157–182.

[HasN] P. Neittaanmäki, J. Haslinger, *Finite Element Approximation for Optimal Shape, Material and Topology Design*, Wiley, 2nd ed., 1996.

[Hei1] M. Heinkenschloss, *Formulation and analysis of a sequential quadratic programming method for the optimal Dirichlet boundary control of Navier-Stokes flow*, in: Hager, William H. (ed.) et al., *Optimal Control: Theory, Algorithms, and Applications.* Proceedings of a conference, University of Florida, Gainesville, 1997. Kluwer Academic Publishers. Appl. Optim. 15, 1998, 178–203.

[Hei2] M. Heinkenschloss, *The numerical solution of a control problem governed by a phase field model*, Optim. Methods Softw. 7, 1997, 211–263.

[Hi] M. Hinze, *Optimal and Instantaneous Control of the Instationary Navier-Stokes Equations*, Habilitation Thesis, Fachbereich Mathematik, Technische Universitt Berlin, 2000.

[HK1] M. Hinze, K. Kunisch, *On suboptimal control strategies for the Navier-Stokes equations*, ESAIM, Proc. 4, 1998, 181–198.

[HK2] M. Hinze, K. Kunisch, *Newton's method for tracking type control of the instationary Navier–Stokes equations*, Preprint, October 1999, to appear in ENUMATH 99, P. Neittaanmki et. al. (eds.)

[IK] K. Ito, K. Kunisch, *Augmented Lagrangian-SQP methods for nonlinear optimal control problems of tracking type*, SIAM J. Contr. Optim. 34, 1996, 874–891.

[JW] A. Jonsson, H. Wallin, *Function Spaces on Subsets of* $\mathbb{R}^n$, Harwood Academic Publishers, Mathematical Reports, vol. 2, 1984.

[JL] A. Jouini, P.G. Lemarié–Rieusset, *Analyses multirésolutions biorthogonales et applications*, Ann. Inst. Henri Poincaré, Anal. Non Lineaire 10, 1993, 453–476.

[Ka] S. Kaczmarz, *Angenäherte Auflösung von Systemen linearer Gleichungen*, Bulletin de l'Academie Polonaise des Sciences et Lettres A35, 1937, 355–357.

[Kf] A. Kauffmann, *Optimal Control of the Solid Fuel Ignition Model*, Ph.D. Thesis, TU Berlin, 1998.

[KP] K. Kunisch, G. Peichl, *Shape optimization for mixed boundary value problems based on an embedding domain method*, Dyn. Contin. Discrete Impulsive Syst. 4, 1998, 439 - 478.

[K1] A. Kunoth, *Multilevel Preconditioning*, Ph.D. Thesis, FU Berlin, Verlag Shaker, Aachen 1994.

[K2] A. Kunoth, *Computing integrals of refinable functions — Documentation of the program*, Version 1.1, Technical Report ISC-95-02-MATH, Institute for Scientific Computation, Texas A&M University, May 1995.

[K3] A. Kunoth, *Multilevel preconditioning — Appending boundary conditions by Lagrange multipliers*, Adv. Comp. Maths. 4, 1995, 145–170.

[KK] A. Kunoth, A. Kurdila, *Wavelet techniques for saddle point problems in optimal control*, Manuscript.

[KPV] A. Kunoth, Piquemal, J. Vorloeper, *Multilevel preconditioners for discretizations of elliptic boundary value problems — Experiences with different software packages*, IGPM-Preprint #194, July 2000.

[LT] L. Levaggi, A. Tabacco, *Wavelets on the interval and related topics*, Preprint #11, Dipartimento di Matematica, Politecnico di Torino, 1997, to appear in Rend. Sem. Univ. Pol. Torino.

[Li] J.L. Lions, *Optimal Control of Systems Governed by Partial Differential Equations*, Springer, Berlin, 1971.

[LM] J.L. Lions, E. Magenes, *Non-Homogeneous Boundary Value Problems and Applications*, Vol. I, Springer, 1972.

[MKM] G.I. Marchuk, Y.A. Kuznetsov, A.M. Matsokin, *Fictitious domain and domain decomposition methods*, Sov. J. Numer. Anal. Modeling 1, 1986, 3–35.

[Ma1] R. Masson, *Biorthogonal spline wavelets on the interval for boundary value problems*, Math. Mod. Meths. in Appl. Sci. 6, 1996, 749–791.

[Ma2] R. Masson, *Wavelet Methods in Numerical Simulation for Elliptic and Saddle Point Problems*, Ph.D. Thesis, Université Paris-VI, January 1999.

[MM] H. Maurer, H.D. Mittelmann, *Optimization techniques for solving elliptic control problems with control and state constraints: Part 1. Boundary Control*, Comput. Optim. Appl. 16, 29–55, 2000.

[Na] F. Natterer, *Algorithms in tomography*, in: *The State of the Art in Numerical Analysis*, I.S. Duff (ed.) et al., Oxford, Clarendon Press, Inst. Math. Appl. Conf. Ser., New Ser. 63, 1997, 503–523.

[NT] P. Neittaanmäki, D. Tiba, *An embedding of domains approach in free boundary problems and optimal design*, SIAM J. Contr. Optim. 33, 1995, 1587–1602.

[Nep] S.V. Nepomnyaschikh, *Ficticious space method on unstructured grids*, East–West J. Numer. Maths. 3, 1995, 71–79.

[Ni] N.K. Nichols, *On the convergence of two–stage iterative processes for solving linear equations*, Siam J. Numer. Anal. 10, 1973, 460–469.

[OC] J.T. Oden, G.C. Carey, *Finite Elements - Mathematical Aspects*, Vol. IV, Prentice–Hall, New Jersey, 1983.

[O1] P. Oswald, *On discrete norm estimates related to multilevel preconditioners in the finite element method*, in: Constructive Theory of Functions, K.G. Ivanov, P. Petrushev, B. Sendov, (eds.), Proc. Int. Conf. Varna 1991, Bulg. Acad. Sci., Sofia, 1992, 203–214.

[PCL] A.I. Pehlivanov, G.F. Carey, R.D. Lazarov, *Least-squares mixed finite elements for second-order elliptic problems*, SIAM J. Numer. Anal. 31, 1994, 1368–1377.

[PRG] L. Petzold, J.B. Rosen, Ph.E. Gill, L.O. Jay, K. Park, *Numerical optimal control of parabolic PDEs using DASOPT* L.T. Biegler et al. (eds.), Large-Scale Optimization with Applications. Part 2: Optimal Design and Dontrol, Springer, IMA Vol. Math. Appl. 93, 1997, 271–299.

[Pi] J. Pitkäranta, *Boundary subspaces for the finite element method with Lagrange multipliers*, Numer. Math. 33, 1979, 273–289.

[PW] W. Proskurowski, O. Widlund, *On the numerical solution of Helmholtz equation by the capacitance matrix method*, Math. Comp. 30, 1979, 433–468.

[Ri1] A. Rieder, *A domain embedding method for Dirichlet problems in arbitrary space dimensions*, RAIRO, Modelisation Math. Anal. Numer. 32, 1998, 405–431.

[Ri2] A. Rieder, *Embedding and a–priori wavelet–adaptivity for Dirichlet problems*, Preprint # 99/9, Institut für Wissenschaftliches Rechnen und Mathematische Modellbildung, Universität Karlsruhe, September 1999.

[Sch] R. Schneider, *Multiskalen- und Wavelet-Matrixkompression: Analysisbasierte Methoden zur effizienten Lösung großer vollbesetzter Gleichungssysteme* (in german), Advances in Numerical Mathematics, Teubner, 1998.

[Sta] G. Starke, *Multilevel boundary functionals for least squares mixed finite element methods*, SIAM J. Numer. Anal. 36, 1999, 1065–1077.

[St1] R. Stenberg, *On some techniques for approximating boundary conditions in the finite element method*, J. Comp. Appl. Maths. 63, 1995, 139–148.

[St] E.M. Stein, *Singular Integrals and Differentiability Properties of Functions*, Princeton University Press, 1970.

[Sw] W. Sweldens, *The lifting scheme: A construction of second generation wavelets*, SIAM J. Math. Anal. 29, 1998, 511–546.

[Ta] K. Tanabe, *Projection method for solving a singular system of linear equations and its applications*, Numer. Math. 17, 1971, 203–214.

[Toi] J. Toivanen, *Ficititious Domain Method Applied to Shape Optimization*, Ph.D. Thesis, University of Jyväskylä, 1997.

[Tr] F. Tröltzsch, *On the Lagrange–Newton–SQP method for the optimal control of semilinear parabolic equations*, SIAM J. Cont. Optim. 38, 1999, 294–312.

[Vo] J. Vorloeper, *Multiskalenvefahren und Gebietszerlegungsmethoden* (in german), Diploma thesis, IGPM, RWTH Aachen, August 1999.

[Z] E. Zeidler, *Nonlinear Functional Analysis and its Applications; III: Variational Methods and Optimization*, Springer, 1985.

Index